101 Group Theory for Chemists

Christoph Sontag PhD

ISBN-13: 978-1533455178

DEDICATION

This booklet is dedicated to my friend and instructor

Associate Professor Dr. Apinpus Rujiwatra
Chiang Mai University.

As expert in crystal structures and symmetry, her teaching helped many students to get familiar with the concept of symmetry in chemistry and their many applications.

Preface

Group Theory is maybe the most anxiety-provoking term in chemical education – it is a mathematical concept that sounds very abstract to most learners. In this little book I try to make this theory come to life and make it very practical – leaving out theoretical concepts that, for a mathematician, seem to be absolutely necessary. But for applications to chemical problems we should focus on just the part that is useful for our purpose.

All drawings were done by the freeware program "ChemSketch 2015" which I highly recommend for all chemistry students and teachers.

 Please follow me on the path to exploring the amazing way how chemical molecules and their properties are related to symmetry behaviors, implicating that there is a more fundamental, mathematical-like law behind nature.

Please read the introduction to each topic carefully and try to answer the following questions indicated by numbers like $\boxed{1}$
You will find the answer in the appendix.

Christoph Sontag, PhD

Phayao, Thailand

October 2016

Contents

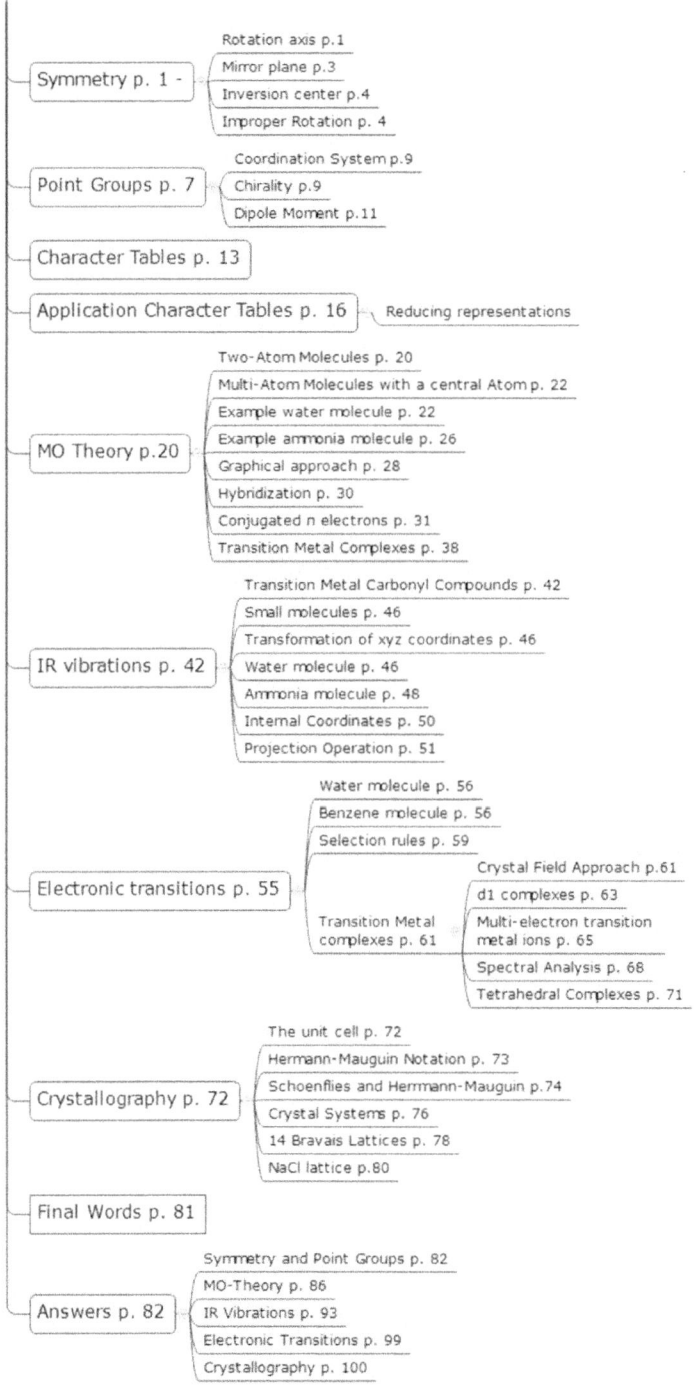

Symmetry

We all know symmetry from our daily observations – often we find it in buildings, drawings and even our own body. Our both hands are like image and mirror-image to each other.

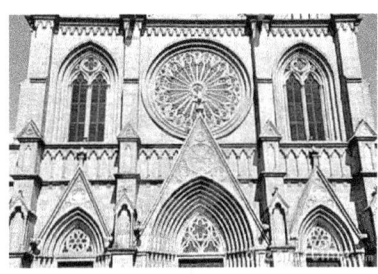

Symmetry conveys order and calmness to our eye. The absence of any symmetry in opposite gives the impression of chaos.

For example in old church buildings, symmetry had been an important element used in many details.

But how can we "quantify" symmetry – how can we say that one thing is "more symmetrical" than another ?

There is a simple way to do that – we just count the number of **symmetry-elements** or symmetry-operations.

There are only three basic elements: rotation ("c"), mirror("σ") and inversion("I"). For completeness there is also the trivial "operation" of identity ("E"):

Identity E

This is an operation that just does nothing – the reason why we consider it, is that combinations of symmetry operations have to lead to an operation within the same point group (see later).

Rotation Symmetry (rotation axis c_n)

When looking at the big round window alone, we can find mirror planes as well as rotation symmetry.

A symmetry-element / operation is defined as any movement that transforms a thing into itself:

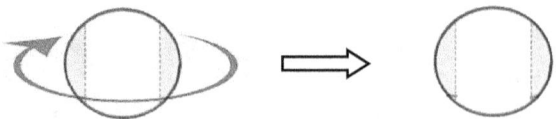

How many different rotation axis are present ? $\boxed{1}$

Rotating a molecule like ammonia around the z-axis by 360°/3 = 120° would give a molecule that cannot be distinguished from the original:

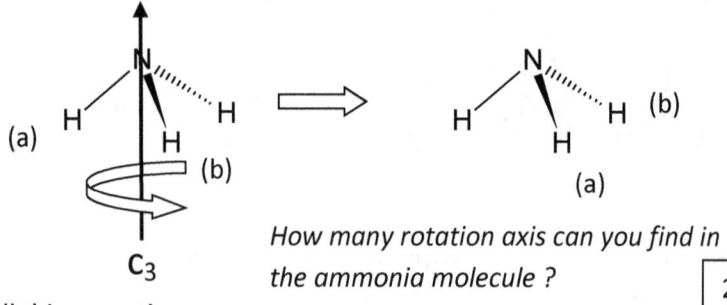

How many rotation axis can you find in the ammonia molecule ? $\boxed{2}$

We call this operation a "c_3" rotation: 3 because we can rotate three times in order to get the same molecule. Or, in other words, we rotate for 360°/n degrees with n=3 and get the same molecule.

Now let's be very precise here:
the molecule on the right looks exactly the same as on the left – if we do not make a distinction between the three hydrogen atoms. If the hydrogen atom in position (a) on the left is equal to the hydrogen (b), then we have rotation symmetry.

A highly symmetric molecule is benzene (notice that the double bonds are de-localized !)

The main rotation axis C_6 points out of the paper plane - it is the main axis because it has the highest n, therefore defining the z-axis of the whole molecule. Notice that there are two sets of 3 c_2 axis in the plane of the molecule *(graphic by: http://www.chemtube3d.com/SymBenzeneD6h.htm)*

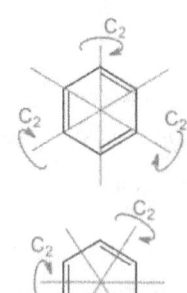

Mirror Symmetry (mirror plane σ)

In the above picture we can immediately identify a mirror plane in the middle of the image.

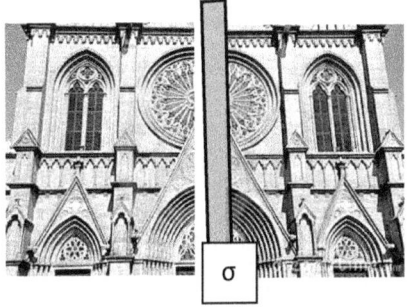

The right half of the image is a mirror of the left half.

Symmetry planes in the ammonia molecule:

The mirror operation would switch the positions of hydrogen atom (a) and (b)

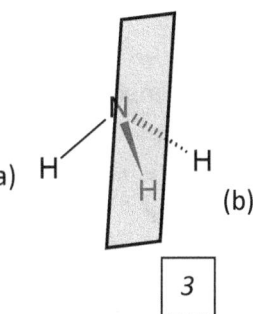

(a)   (b)

How many mirror planes can you identify in the ammonia molecule ?

3

3 kinds of mirror planes

Mirror planes which include the rotation axis with the highest n are called "vertical" (since this axis defines the z-axis of the molecule) or σ_v.

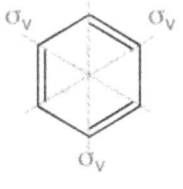

For a benzene ring, the main c-axis would be c_6 (coming out of the paper plane) and we can find three σ_v planes.

A mirror plane perpendicular to the c-axis is called "horizontal" or σ_h.

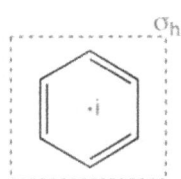

In our example of a benzene molecule, this would be the mirror plane which is identical to the molecule plane itself.

Finally a mirror plane which divides atoms (instead of containing them) is called "diagonal" or σ_d

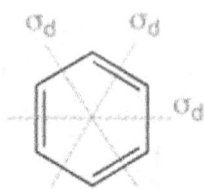

It is not always possible to assign one of these symbols.

Inversion center i

An inversion center is a point through which each part of a molecule can be mirrored and get the same molecule.

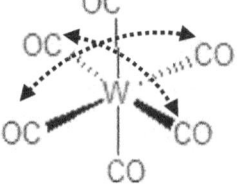

This operation is quite common for transition metal compounds, for example the metal in the center is the inversion center:
each CO ligand will be transformed to the other one on the opposite side.
(notice also that all d-orbitals have an inversion center)

Screw Operation S_n

Besides the three obvious symmetry elements, there is by definition a fourth one, which is actually a combination of rotation and mirror operation: the screw-operation or "improper rotation", called S_n.

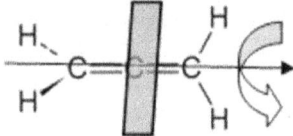

An example is propadiene: combine a c_2 operation with a mirror operation:

Thereby the order does not make a difference

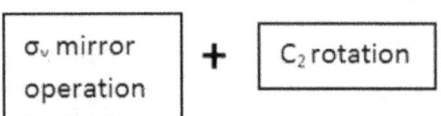

This symmetry element is quite rare in small molecules but can become prominent in macromolecules (the famous double helix structure of

DNA) and crystal structures.

Order of symmetry N

To find the order of symmetry of a molecule, we have to count all possible symmetry operations.

But we must be careful here:
for the ammonia molecule we have to count <u>two</u> rotation operations – a rotation by 120° <u>and</u> by 240° will turn the molecule in its identical picture !

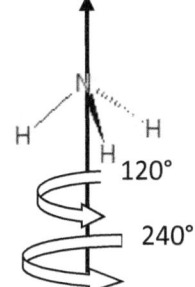

In addition we would have three different mirror planes which all include the z-axis and one of the three hydrogen atoms:

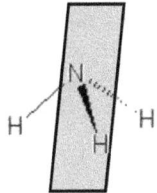

The total of operations for ammonia would then be:

h = 1* identity (<u>always</u>) + 2* rotation + 3* mirror = 6

The obvious operation "identity" looks obsolete but from a mathematical point of view it is necessary, so all things have at least one symmetry element.

Extension of the symmetry concept

The symmetry concept is not limited on atoms in a molecule – symmetry is a universal concept for all "things", for atoms, bonds, orbitals and even for movements.

An important application is in MO theory: atomic and molecule orbitals as combinations always have distinct symmetries. And only orbitals with the same symmetry can interact and form a bond !

For example consider the d_{yz} orbital of a metal and the p_z orbital of a halogen ion:

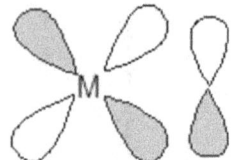

The metal-d and the halogen-p orbital are both anti-symmetric with respect to a mirror plane – or: they both have an inversion center in the atom position !

Practice: which symmetry elements can you find for:

4

Acetone:

a

H₃C ⟍ ⟋ CH₃ (with O double-bonded at top)

Methyl- cyclopentane:

CH₃

b

Ethanediol:

Ironpentacarbonyl:

OC ⟍ Fe —CO
OC ⟋ |
CO (with CO at top and bottom)

c

Ehanediol:

H, H⟍ OH
⟋ ⟍ ""H
HO H

d

You can visualize symmetry operations using the online tool in www.molwave.com: go to 3DMolSym which opens a browser window (you need to install the Flash-plugin for your browser), where you can select different molecules to check about their symmetries.

Draw the following molecule after performing different symmetry operations (mark the 4 water ligands by their numbers):

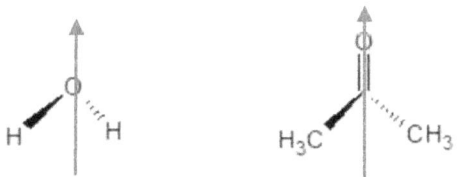

Point Groups

Molecules that have the same symmetry elements belong to the same "family". For example the water and acetone molecule both have one c_2 axis and two mirror planes:

Both molecules have four symmetry elements:

- one C_2 axis

- two mirror planes σ

- one identity operation E

At first the identity operation seems obsolete - but a point group is characterized by the fact that <u>each combination of two symmetry operations must result into an operation which is also a member of the group</u>. For example to apply a rotation c_2 two times results in the identity operation ("do nothing").

This "family" gets the name "c_{2v}" - this name or "point group" can be determined by a flow diagram, in which we identify stepwise which elements are in a molecule.

Apart from special cases (linear molecules and highly symmetric octahedron and tetrahedron) we can identify two big groups, the C- and the D- group. C-groups have a main rotation axis and "vertical" mirror plane(s) like in the examples above - D-groups in addition have a horizontal mirror plane.

<u>Linear:</u> **D_{∞h}** for A-B-A   (with inversion center *i*)

         C_{∞h} for A-B

<u>Tetrahedral T_d:</u>

<u>Octahedral O_h:</u>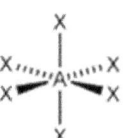

<u>All other symmetries:</u>

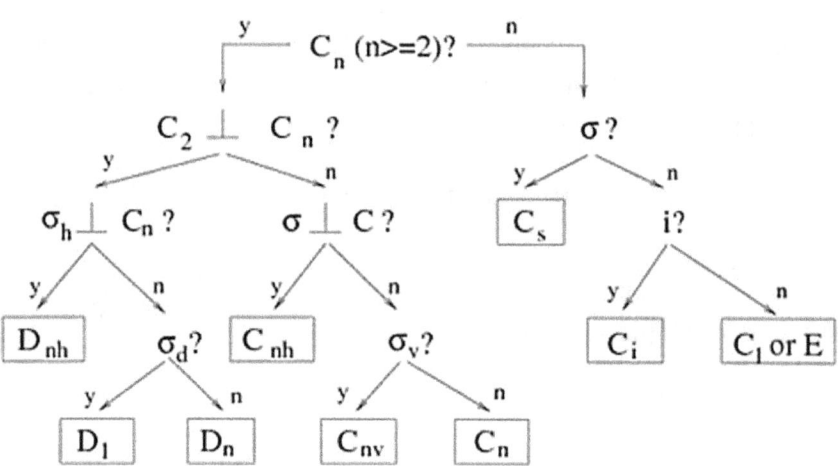

(http://en.wikibooks.org/wiki/Introduction_to_Mathematical_Physics/N_body
_problem_in_quantum_mechanics/Molecules)

*Practice online at: **http://symmetry.otterbein.edu/challenge/***

Determine the point group of the following molecules:

6

SF₅Cl   CCl₄   H₂C=CH₂   H₂C=CF₂   SO₄²⁻   SO₃

PCl_5 PCl_3 $O=PF_3$ $(PPh_3)_3RhCl$ mer-$(NH_3)_3(Cl)_3Co$

Coordination System

The coordination system can be defined by: the main rotation axis is the z-axis, in our example the C_2 axis. The other two axes can be found with the so-called "right hand rule":

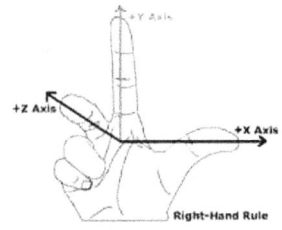

Chirality (optical activity)

A chiral molecule has two isomer forms which are not superimposable - they behave like image and mirror-image. Left and right hand is a classic example (thus the name "chiral' coming from the word "hand").

These two isomers are called "enantiomers".

Their chemical properties are identical; therefore they cannot be separated by chemical methods.

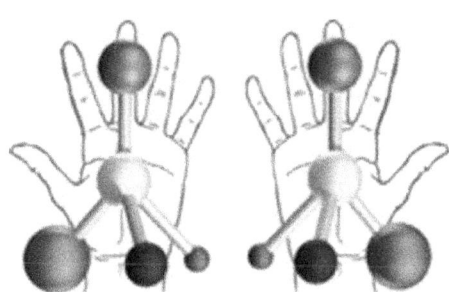

From the view of symmetry, a chiral molecule cannot have a mirror plane or an inversion center.

Chiral molecules can NOT have an improper rotation axis s_n .
Since $s_1 = \sigma$, a molecule with any mirror plane cannot be chiral.

Draw the s_3 axis into the picture of an "optically active" carbon center, which converts this enantiomer into its mirror image. (HC*FClBr)

7

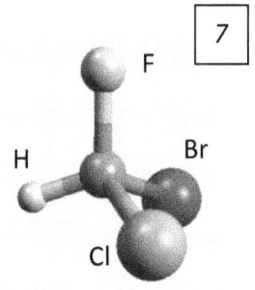

(Therefore the molecule has no s_n axis, because the resulting molecule is not the same as before !)

9

This configuration is called "R" form: we look to the center of a molecule or ligand, so that the atom <u>with the lowest atomic number</u> is behind (or the rest of the molecule):

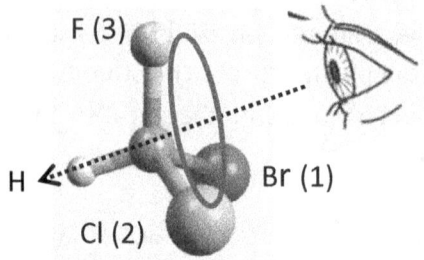

F (3)

H ◄

Cl (2)

Br (1)

Going from Br to Cl and to F, we have to move clockwise = called "R"

The other direction is anti-clockwise and called "S"

(for an extended explanation see for example:

http://www.ochempal.org/index.php/alphabetical/q-r/rs-convention/)

Suggest which of the following molecules is chiral:

8

Molecule 1:

CH_3

H—C—H

Br

Molecule 2:

CH_3

H—C—$\overset{H_2}{C}$—CH_3

CH_2

CH_2

CH_3

Molecule 3:

H—C(=O)—CH_3

Molecule 4:

H—C(=O)

HC—OH

OH—CH

HC—OH

HC—OH

CH_2OH

Indicate also the chiral carbons (if any)

10

Dipole moments

A dipole moment in a molecule arises from an uneven distribution of partial charges inside a molecule.

It can be found by assigning dipole moments to each bond in a molecule (arrows going from an atom with lower electronegativity to another with higher) and add all arrows together as vectors.

Compare the molecular dipole moments of water and carbon dioxide:

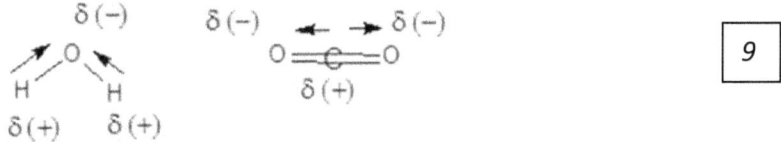

9

> *A molecule can only have a dipole moment if there is a displacement of charge along the x,y or z-axis.*
> *Molecules with any horizontal mirror plane, cannot have a dipole moment.*

For example in the linear CO_2 molecule, the z-axis would be the axis of the molecule - there is no change in polarity because the two small dipole moments compensate each other.

When a molecule has a dipole moment, it is also "polar" - meaning: we can conclude about the solvent property of a molecule from this property.

Examples: *Order the following solvents according to increasing polarity:*

10

CH_3CH_2-OH *(ethanol)* $H_3C=C=O$-CH_3 *(acetone)* CCl_4 *(tetra)*

CH_3-Cl *(chloroform)* C_6H_5-CH_3 *(toluene)* $(CH_3)_2$ S=O *(DMSO)*

To know about polarity is an essential tool for all kinds of **chromatography** (TLC, HPLC, column chromatography)

Which of the above solvents would be suitable for a TLC of :

11

a	b	c

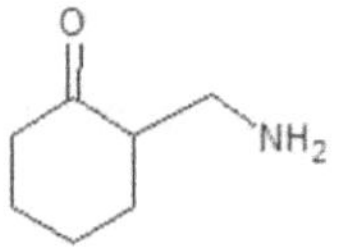

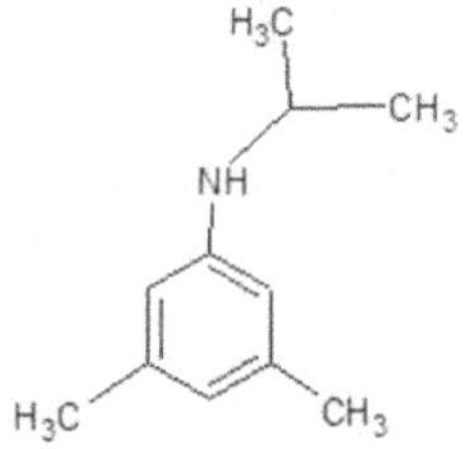

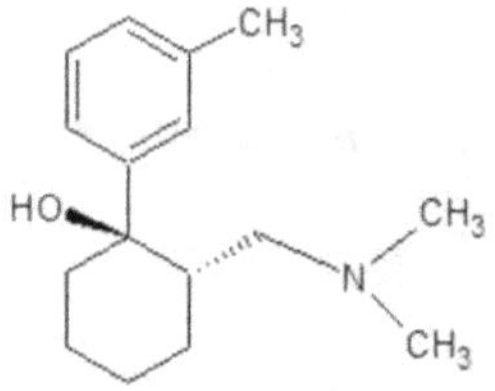

Besides these practical consequences, the dipole moment is the cause for the absorption of infrared-light by vibrations: an electromagnetic wave can cause bonds to vibrate, if and only if this vibration movement causes a change in dipole moment.

Based on this information, decide which bond vibration causes such a change in dipole moment of the carbonate molecule:

12

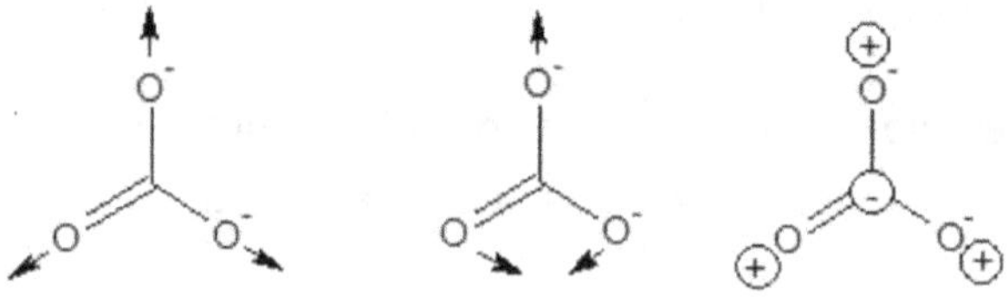

Character Tables

A molecule (or in general a "thing") that has a certain set of symmetry elements belongs to a specific point group. Inside this group, a certain set of possible transformations can be identified.

Interestingly the number of possible transformations is the same as the number of possible symmetry-elements: this leads to a so-called character table of this group.

For the above mentioned c_{2v} group for a water molecule, this table looks like this: (http://www.webqc.org/symmetrypointgroup-c2v.html)

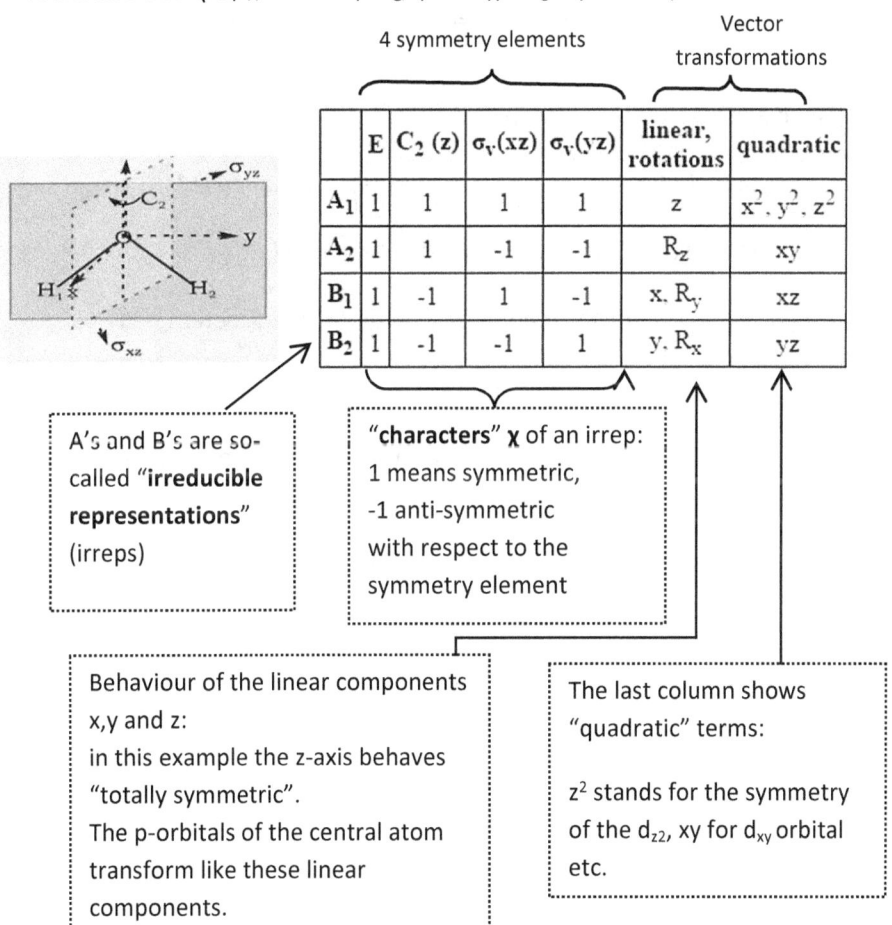

4 symmetry elements

Vector transformations

	E	C_2 (z)	σ_v(xz)	σ_v(yz)	linear, rotations	quadratic
A_1	1	1	1	1	z	x^2, y^2, z^2
A_2	1	1	-1	-1	R_z	xy
B_1	1	-1	1	-1	x, R_y	xz
B_2	1	-1	-1	1	y, R_x	yz

A's and B's are so-called "**irreducible representations**" (irreps)

"**characters**" χ of an irrep:
1 means symmetric,
-1 anti-symmetric
with respect to the symmetry element

Behaviour of the linear components x,y and z:
in this example the z-axis behaves "totally symmetric".
The p-orbitals of the central atom transform like these linear components.

The last column shows "quadratic" terms:

z^2 stands for the symmetry of the d_{z2}, xy for d_{xy} orbital etc.

Let's have a closer look at the y-vector:

- identity E (what is true for all elements): no change (+1)

- σ_{yz} mirror plane (the paper plane): no change (+1)

- σ_{xz} mirror plane, (perpendicular): reverses the direction (-1)

- C_2 rotation by 180°: reverses also the direction (-1)

Because of this behaviour, the y-axis vector has the "character" of B_2, indicated by the "y" in the next column.

The x,y and z vectors have the same behaviour as the p_x, p_y and p_z orbital - therefore orbitals are often named after their characters.

For the p_y-orbital of oxygen in a water molecule, we would describe it as "b_2"

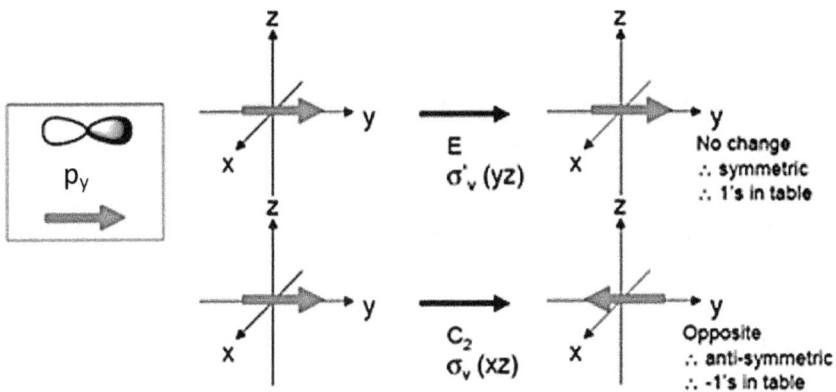

We can see not only the characters of the 3 coordination vectors, x,y and z, but also combination terms in the last column, like xy etc.

These correspond to d-orbitals:
(don't get confused about the term x^2, it describes the $d_{x^2-y^2}$ orbital)

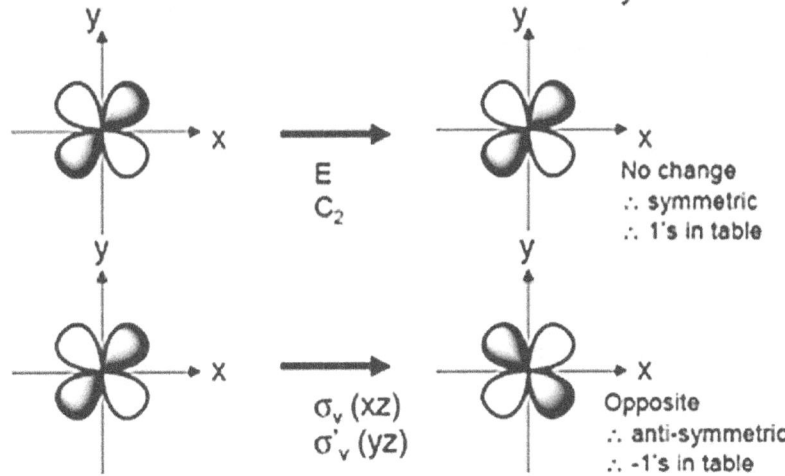

So we can see, that the symmetry elements (or point group) describe the symmetry of the <u>whole molecule</u>, and the characters describe the symmetry of <u>parts of a molecule</u> (in this example orbitals) !

The representations (rows in the character tables) are basic symmetries that are "hidden" inside the specific point group. They are called "irreducible representations" ("irreps").

Applications of Character Tables

The information that is included in character tables is very useful to find the symmetry behaviour of molecular properties - such as bonding (described by orbitals) and spectroscopic terms (vibrations and electronic transitions). This reflects the deep mathematical nature behind creation.

We can follow a general scheme to analyze these properties:

① Define the basis set:

which is a set of properties for the molecule under consideration - that can be the atomic orbitals of a central atom which form bonds and the group of orbitals around the central atom ("group orbitals").

Another possible set to look at are stretching and bending vibrations, expressed by atom dislocations or vibration vectors.

Then we have to find the characters of this basis set in the point group of the molecule.

For example the two bonds in a water molecule as part of the molecule can be described by a "**representation**" Γ.
We note the number of <u>unshifted</u> bonds

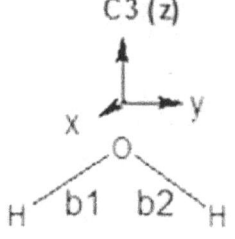

C_{2v}	E	C_2	σ_{xz}	σ_{yz}
Γ_{b1b2}	2	0	0	2

A rotation around the z-axis exchanges bond1 and bond2, so the character is 0. The same happens by mirror at the xz plane. A mirror plane in the molecule plane would leave both bonds unchanged, therefore the character 2.

②Reduction of representations

Parts of a molecule behave like representations which can be "reduced" to the basic representations or irreps.

Γ_{b1b2} can be "reduced" to the basic representations:

Γ_{b1b2}: 2 0 0 2

A_1 : 1 1 1 1

B_2 : 1 -1 -1 1

We have to find out, which of the "irreducible representations" are "inside" Γ_{b1b2} by some algorithm. We ask:

How many times is A_1 included in Γ_{b1b2} ?

Answer: $n = 1/h * [\Sigma (\text{# symm op}) * \chi(A_1) * \chi(\Gamma_{b1b2})]$

C_{2v}	1 E	1 C_2	1 σ_{xz}	1 σ_{yz}	h = 4
A_1	1	1	1	1	
Γ_{b1b2}	2	0	0	2	
contribution	1 * 1 * 2 = 2	1 * 1 * 0 = 0	1 * 1 * 0 = 0	1 * 1 * 2 = ?	
Sum:	$(1*1*2) + (1*1*0) + (1*1*0) + (1*1*2)$				= 4

This scary formula becomes very simple when we look at the table above: we multiply the number of the symmetry element (in this case only 1, not explicitly written) with the character of A_1 in this column (1) and with the character of our Γ_{b1b2} : $(1*1*2)$

We do this in each column, finally add all values together, and divide by h gives:
$n(A_1) = 1/4 * [(1*1*2) + (1*1*0) + (1*1*0) + (1*1*2)] = 1/4 * 4 = 1$

That means that A_1 is <u>one time included</u> in Γ_{b1b2}.

When you calculated correctly, you should find that also B_2 is included in it, but not A_2 or B_1.

We can "see" this quite easily by just looking at the character table:

the combination 2 0 0 2 can be found by addition of the rows A_1 and B_2:

C_{2v}	E	C_2	σ_{xz}	σ_{yz}
A_1	1	1	1	1
B_2	1	-1	-1	1
$\Gamma_{b1b2} = A_1 + B_2$	2	0	0	2

In simple cases like this, we do not have to go through the tedious process to apply the reduction formula, but can find the combination by try-and-error from the character table directly - just add to the first row another row and see if it fits with your combination.

We know now that the two O-H bonds in water have A_1 and B_2 representations. From the character table we see that the z-vector behaves like A_1 and the y-vector like B_2. (the x-component which has B_1 symmetry does not participate in the bonds). In other words: the p_y and p_z orbitals of oxygen are involved in the bonding but not the p_x orbital.

③ Projection Operators

To know how these representations look like, we use a method called "projection": we "project" the entity of question (an orbital, a bond or a vibration arrow) onto the irrep (in this case A_1 and B_2) and call it P(..):

C_{2v}	E	C_2	σ_{xz}	σ_{yz}	
A_1	1	1	1	1	
b1 becomes P(b1)* =	b1	b2	b2	b1	
A_1 x P(b1) =	b1	b2	b2	b1	= 2 (b1 + b2)
B_2	1	-1	-1	1	
B_2 x P(b1) =	b1	- b2	- b2	b1	= 2 (b1 - b2)

* P(b1) means projection of the bond b1 onto the symmetry elements

Apart from the factor 2 (which can be normalized), the important information is that the A_1 symmetry belongs to a combination of both bonds, whereas the B_2 symmetry to a combination of b1 with the negative of b2.

If we look at the bonds as orbitals, these molecular orbitals should look like:

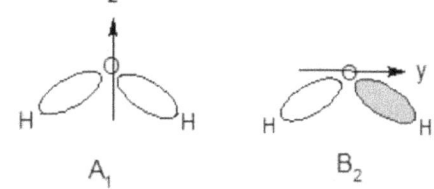

In terms of atomic orbitals (AO) of the oxygen, we can already see that one bond is made from the $2p_z$ AO of oxygen and the other one from the $2p_y$ AO.

If we look at the bonds as stretching vibrations, then the two irreps A_1 and B_2 stand for two molecular vibrations:

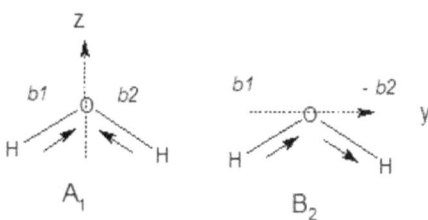

We discuss the construction of orbitals and vibration motions in the following sections.

Important for now is to understand that we can describe different properties of a molecule (orbitals, vibrations) with the same symmetry concept.

Applications in MO Theory

As we have seen in the previous part, each orbital in a molecule can be described by its characters.

The importance of that lies in the fact that only orbitals with the same character can overlap, means: can form a bond.

Two-Atom Molecules

Take as first example the O_2 molecule:

The two p_x orbitals of each oxygen can form a so-called σ-bond: (careful - this "σ" does not stand for mirror plane but describes a single bond which is symmetric about rotation)

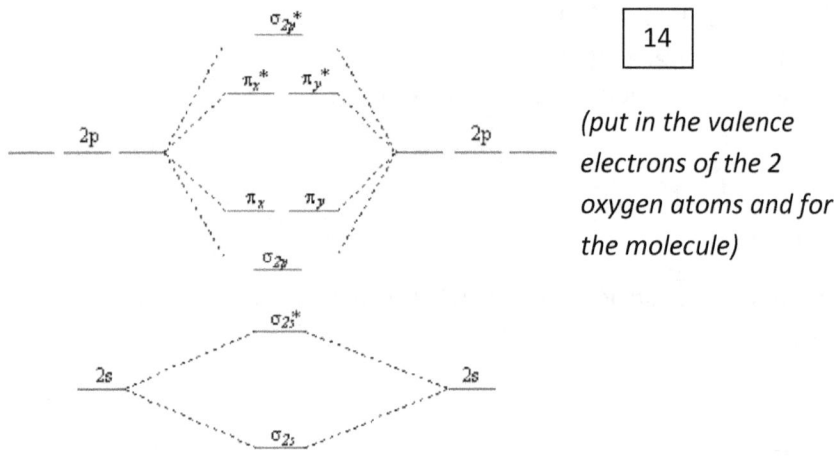

14

(put in the valence electrons of the 2 oxygen atoms and for the molecule)

Notice that 3 "bonds" are formed: between two p_x orbitals a "σ-bond" (rotation symmetry around the bond)

and two times between two p_y and p_z orbitals (anti-symmetric for the rotation around the bond)

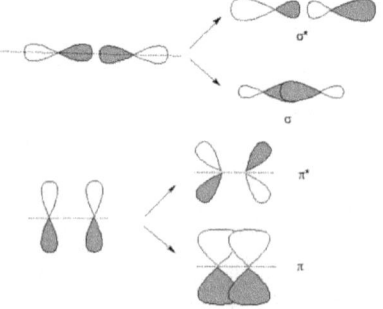

The resulting O_2 molecule has $C_{\infty v}$ symmetry with the character table:

	E	$2C_\infty$	...	$\infty \sigma_v$	linear, rotations	quadratic	
$A_1=\Sigma^+$	1	1	...	1	z	x^2+y^2, z^2	
$A_2=\Sigma^-$	1	1	...	-1	R_z		
$E_1=\Pi$	2	$2\cos(\varphi)$	...	0	(x, y) (R_x, R_y)	(xz, yz)	
$E_2=\Delta$	2	$2\cos(2\varphi)$	...	0		(x^2-y^2, xy)	
$E_3=\Phi$	2	$2\cos(3\varphi)$	...	0			
...	...		...				

We can identify the z-vector (= p_z orbitals) with A_1 symmetry, so that 2 p_z orbitals can interact (forming a σ-bond).

The p_x and p_y orbitals have the same symmetry (E_1) and therefore can overlap, forming a degenerated set (same energy) of two π-bonds.

We have to remember that s-orbitals are always totally symmetric with A_1 symmetry. This leads to an additional interaction between the 2s- with the $2p_z$-orbitals what is pronounced for N_2. But for O_2 the higher electronegativity results in a big energy gap between the 2s and 2p orbitals so this "mix" effect between s- and p-orbitals is low in this case.

For example a N_2 molecule would have some sp-hybridization = mix of s- and p_z-character:

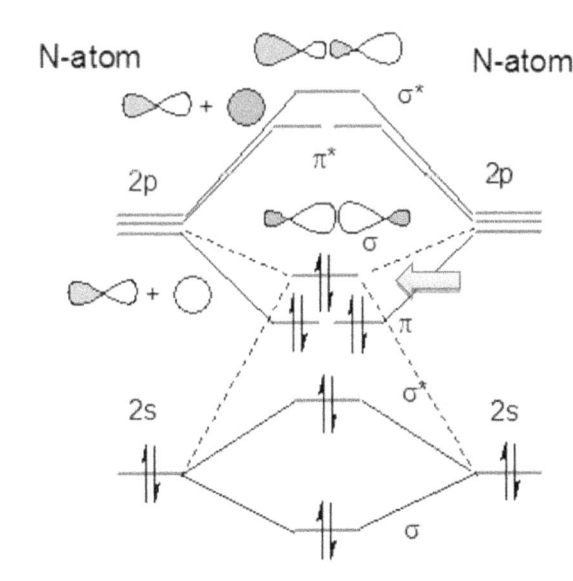

21

> To identify which atomic orbitals (AOs) of one atom can interact with another set of AOs of a second atom, we have to combine those with the same symmetry (in the point group of the resulting molecule).

Construct an MO diagram of an NO molecule.

(remember that O has a higher electronegativity than N, therefore its AO's lie higher in energy. Also consider the s-p mixing effect)

15

Multi-Atom Molecules with a Central Atom - Ligand Group orbitals ("LGO")

The general way to find orbital interactions (= bonds) in a molecule is:

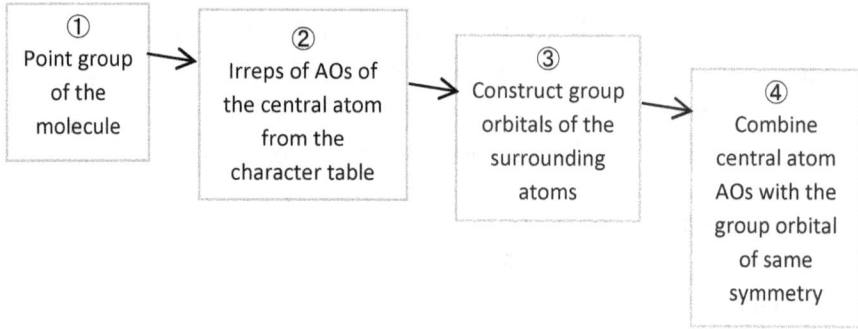

Example 1: Water Molecule

When we want to describe the bonds in a water molecule for example, we first define the coordination system that we want to use. Since water has a c_2 axis, this axis defines the z-direction.

For the definition of the x- and y- axis we should use the so-called "right-hand rule" (see above)

(an intensive discussion about the MO's of the water molecule, see: http://www.huntresearchgroup.org.uk/teaching/teaching_MOs_year2/L2_Not es_2015_web.pdf)

① First we have to find the point group - which is C_{2v}.

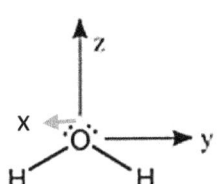

C_{2v}	E	C_2	$\sigma_v(xz)$	$\sigma_v'(yz)$		
A_1	1	1	1	1	z	x^2, y^2, z^2
A_2	1	1	-1	-1	R_z	xy
B_1	1	-1	1	-1	x, R_y	xz
B_2	1	-1	-1	1	y, R_x	yz

② From the character table we see that the oxygen AOs transform as:

2s and $2p_z$ -> A_1 (s orbitals are always totally symmetric)
$2p_x$ -> B_1
$2p_y$ -> B_2

The two hydrogen atoms have each only 1s orbitals, which we have to combine together - we have to form so-called "group orbitals" or "LGO" ("Ligand Group Orbitals")

C_{2v}	E	C_2	$\sigma_v(xz)$	$\sigma_v(yz)$	
A_1	1	1	1	1	z
A_2	1	1	-1	-1	
B_1	1	-1	1	-1	x
B_2	1	-1	-1	1	y
Γ_{2H}	2	0	0	2	

The two hydrogen atoms are un-shifted by the E and the mirror on yz, but change position under C_2 and mirror on xz

By inspecting the rows in the character table we find that the resulting representation Γ_{2H} consists of: $\Gamma_{2H} = A_1 + B_2$

③ To find out how these irreps A_1 and B_2 look like, we have to use our "projection method": check how the atoms under consideration (H1 and H2) behave under the symmetry operations and multiply them with the characters of A_1 and B_2 respectively:

23

C_{2v}	E	C_2	$\sigma_v(xz)$	$\sigma_v(yz)$	
A_1	1	1	1	1	
H1->	H1	H2	H2	H1	
$A_1 \times$ H1:	H1	H2	H2	H1	= 2 (H1 + H2)
B_2	1	-1	-1	1	
$B_2 \times$ H1:	H1	-H2	-H2	H1	= 2 (H1 - H2)

Therefore two combinations of hydrogen 1s-orbitals look like:

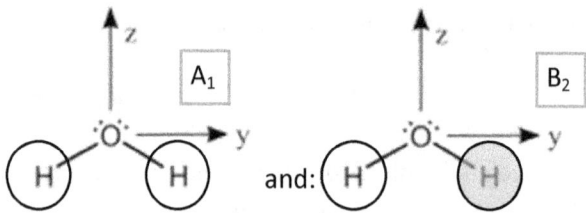

and:

(By convention, a positive phase is in white color, a negative in dark or black)

④ To construct the bonds (= overlapping AO's) we have to combine oxygen AOs with the two hydrogen AO combinations.

(a) The $2p_z$ AO of oxygen has the same symmetry as the symmetric combination of hydrogen AO's (a_1):

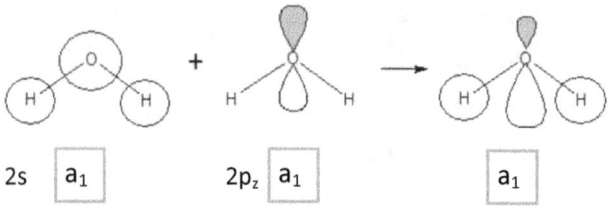

2s a_1 $2p_z$ a_1 a_1

(b) The $2p_y$ AO of oxygen is anti-symmetric with respect to rotation and mirror plane (B_2) and matches with the B_2 combination of hydrogen AOs:

$2p_y$

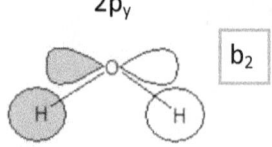

(c) The other combination of the 2s with the $2p_z$ AOs of the oxygen results in a nearly non-bonding MO with a_1 symmetry:

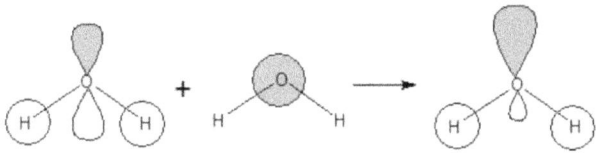

(d) Finally the $2p_x$ AO of oxygen does not match with any combination of the hydrogen AOs - therefore remains non-bonding:

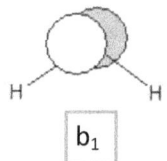

The interaction diagram is (simplified):

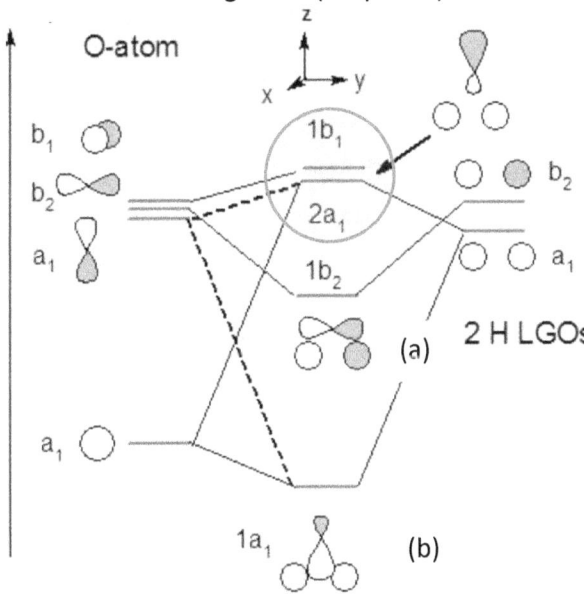

$1a_1$ and $1b_2$ are bonding, $2a_1$ and $1b_1$ non-bonding which corresponds to the two lone pairs on oxygen (which are actually not exactly the same as we expected from the Lewis formula).

The two σ O-H bonds are formed mainly by two interactions:

a) the $1b_1$ MO where 2s and $2p_x$ of Oxygen overlap with the anti-symmetric H_2 combination "σ*" and

b) The $1a_1$ MO where 2s and $2p_z$ of Oxygen interact with the symmetric combination of the H_2 "σ"

Example 2: Ammonia molecule

① determine the point group: C_{3v}

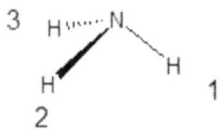

C_{3v}	E	$2C_3$	$3\sigma_v$	
A_1	1	1	1	z
A_2	1	1	-1	
E	2	-1	0	(x,y)

② Representation of the three H atoms (s-AOs)

Under a 120° rotation (C_3), all H atoms change their place (character 0).

Only one mirror operation leaves one H atom at its place (character 1).

C_{3v}	E	$2C_3$	$3\sigma_v$
Γ_{3H}	3	0	1

Reduction of Γ_{3H} yields: **A1 + E**
(which can be seen directly from the character table or by applying the reduction formula)

③ construct three group orbitals of 3 H atoms by projection

See how one hydrogen atom would transform under C_{3v}:

C_{3v}	E	C_3	C_3^2	$\sigma_v(1)$	$\sigma_v(2)$	$\sigma_v(3)$
A_1	1	1	1	1	1	1
A_2	1	1	1	-1	-1	-1
E	2	-1	-1	0	0	0

26

H_1 ->	H_1	H_2	H_3	H_1	H_2	H_3

The projection of one hydrogen atom gives the combination under A_1:

$P_{A1}(H1) = 2H_1 + 2H_2 + 2H_3$ and normalized:
$$= 1/\sqrt{(4+4+4)} \, (2H_1 + 2H_2 + 2H_3) = \mathbf{1/\sqrt{3} \, (H_1 + H_2 + H_3)}$$

And for E symmetry:

C_{3v}	E	C_3	C_3^2	$\sigma_v(1)$	$\sigma_v(2)$	$\sigma_v(3)$
E	2	-1	-1	0	0	0
H_1 ->	H_1	H_2	H_3	H_1	H_2	H_3
H_1 x E	$2H_1$	$-H_2$	$-H_3$	0	0	0

$P_E(1) = 2H_1 - H_2 - H_3 \rightarrow \mathbf{1/\sqrt{6} \, (2H_1 - H_2 - H_3)}$

Analogue we can find two more projections for H_2 and H_3.

$P(H_2) = 2H_2 - H_1 - H_3$
$P(H_3) = 2H_3 - H_1 - H_2$

But E symmetry means that only two sets are necessary - in this case we have to subtract: $P(H_2) - P(H_3) = 3H_2 - 3H_3 \rightarrow \mathbf{P_E(2) = 1/\sqrt{3} \, (H_2 - H_3)}$

Now we can draw these combinations:

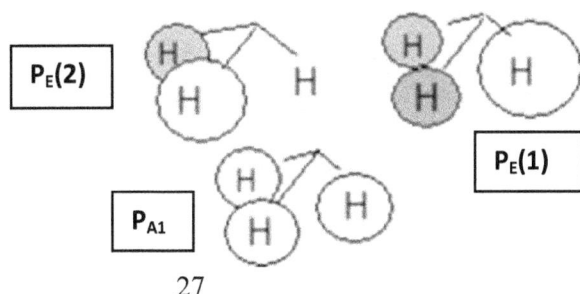

This projection method can become quite complex – fortunately there is a graphical shortcut to construct group orbitals !

Graphical Approach for Group Orbitals

From the atomic orbitals we should remember that the number of nodes (changes in phase) corresponds directly with the energy of an orbital:

an s-orbital has no nodes, p has 1 node, d has 2 and so on.

This concept is valid for group orbitals as well:

First we combine all hydrogen 1s AOs
with the same phase (view from top):

Second we draw a node plane between the three hydrogen AOs - there are two ways to do that:

Notice that the size of the 1s AOs reflects the 1:1 ratio ! The two negative 1s AOs together are as "big" as the one with positive phase.

When we can draw two forms with one node plane, these two are energetically equivalent. We also know that these degenerate group orbitals have higher energy than the totally symmetric one.

④ combine our three LGOs with the AOs of nitrogen:

We can distinguish between the combination of the symmetric LGO (A_1)

with the 2s- and $2p_z$ AO of nitrogen (which are also symmetric towards rotation C_3):

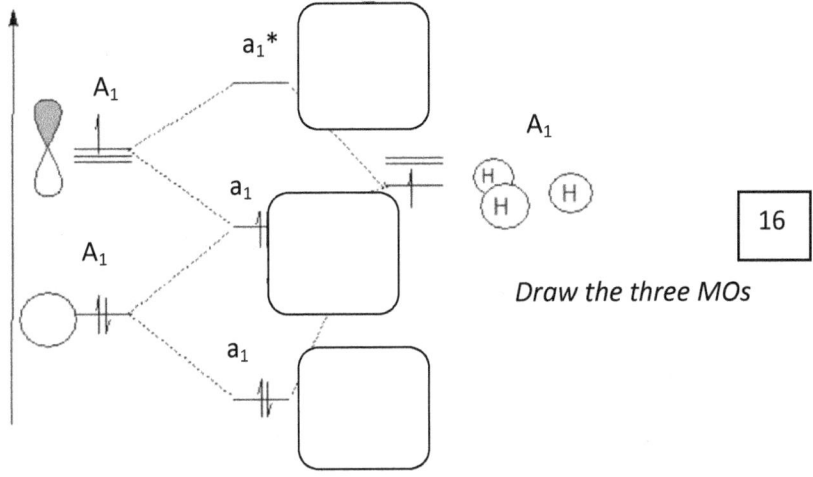

A_1

16

Draw the three MOs

The second interaction is between the other two LGOs with the $2p_x$ and $2p_y$ AO of nitrogen:

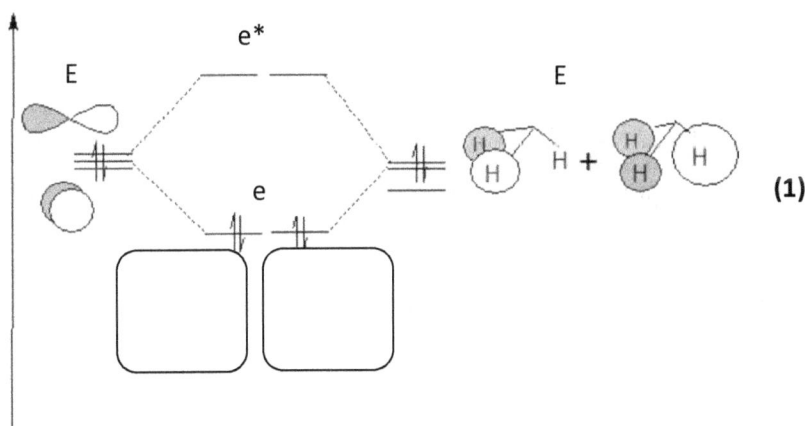

(1)

Draw the 2 bonding combinations

Now draw the complete MO-diagram (combine both interactions) including the electron configuration.

29

Hybridization

This term is used to describe σ-bonds in molecules, which are formed by a mix of s-, p- and d-orbitals of the central atom.

We can verify the pictorial approach by group theory.

For example the three σ-bonds in BX_3 (X = H or Halogen) can be described as follows:

From the Lewis formula we know that these molecules are trigonal-planar. Therefore their point group is D_{3h}.

Character table for D_{3h} point group

	E	$2C_3$	$3C'_2$	σ_h	$2S_3$	$3\sigma_v$	linear, rotations	quadratic
A'_1	1	1	1	1	1	1		x^2+y^2, z^2
A'_2	1	1	-1	1	1	-1	R_z	
E'	2	-1	0	2	-1	0	(x, y)	(x^2-y^2, xy)
A''_1	1	1	1	-1	-1	-1		
A''_2	1	1	-1	-1	-1	1	z	
E''	2	-1	0	-2	1	0	(R_x, R_y)	(xz, yz)

The next step is to find the characters for the 3 bonds under these symmetry operations:

D_{3h}	E	$2C_3$	$2C_2$	σ_h	$2S_3$	$3\sigma_v$
Γ_σ	3	0	1	3	0	1

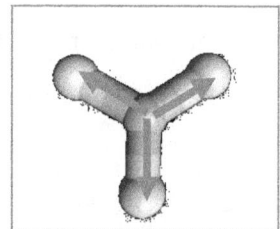

Reduce Γ_σ in the D_{3h} symmetry

When you are right, then Γ consists of: $\Gamma_\sigma = A'_1 + E'$

From the character table we see that E' is also the representation of the 2px and 2py orbital of the central atom. In addition the total symmetric representation in each character table corresponds to the symmetry of an s-orbital.

Therefore we can conclude that Γ consists of an s- and $p_x + p_y$ orbital = **sp² hybrid**

Example: Methane molecule

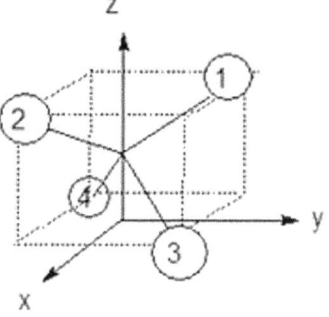

T_d	E	$8C_3$	$3C_2$	$6S_4$	$6\sigma_d$
Γ_{C-H}	4	1	0	0	2

Reducing: $\Gamma_{C-H} = A_1 + T_2$

Which correspond to the central s-AO (A_1) and the 3 p-AOs (T_2), resulting in an **sp³ hybrid.**

17

Find out the hybridization of the central P atom in PCl₃ and PCl₅.
(here you have to look after the representations of d-orbitals as well,
which are z^2, x^2-y^2 and xy, xz, yz (or mixtures of these))

Conjugated π electrons - molecules without a central atom

This concept is prominent for unsaturated and aromatic molecules, describing the π electrons. The idea is to focus only on the π electrons in a molecule since these electrons mostly determine the chemical properties. This concept can be used on all kinds of molecules, inorganic and organic.

Example 1: π- orbitals of the
Nitrite ion NO_2^-

As we can see, this molecule is resonance stabilized - or i.a.w. the π-electrons can freely move over all atoms. This is true for the two lone-pair electrons on O and the two electrons in the double bond - altogether we get 4 π-electrons and 3 MOs, because we have three atoms with p_z orbitals:

(1) we find the point group as C_{2v}

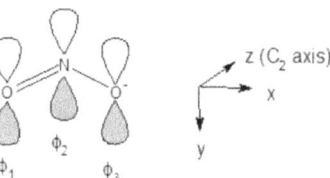

31

(2) we consider how the 3 orbitals transform in this point group:

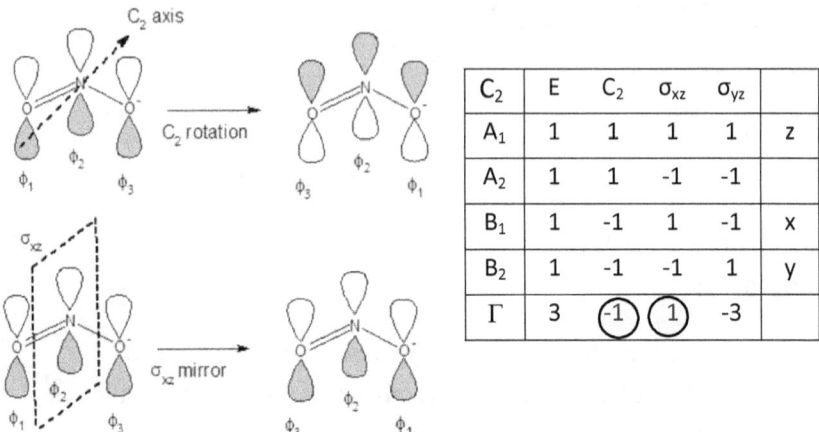

C_2	E	C_2	σ_{xz}	σ_{yz}	
A_1	1	1	1	1	z
A_2	1	1	-1	-1	
B_1	1	-1	1	-1	x
B_2	1	-1	-1	1	y
Γ	3	(-1)	(1)	-3	

Through rotation C_2, ϕ_1 and ϕ_3 change their position and ϕ_2 is rotated but stays at the same place => the character becomes -1

The vertical mirror operation σ_{xz} also interchanges ϕ_1 and ϕ_3 but leaves ϕ_2 as it was before, therefore the character +1.

The other mirror plane σ_{yz} reverses all 3 orbitals, so the character becomes -3.

Reducing Γ leads to: $\Gamma = A_2 + 2\, B_1$

How do these orbital combinations look like ? We apply the mentioned **projection method** : we check how the orbitals ϕ_1 and ϕ_2 behave under the symmetry operations for A_2 and B_1 symmetry.
(Since ϕ_1 and ϕ_3 are equivalent, we need to check only ϕ_1 and $\phi_{2)}$

Step 1:
In the character table we write the characters of the symmetry we want to analyze, in this case first A_2:

Step 2:

C_{2v}	E	C_2	σ_{xz}	σ_{yz}
A_2	1	1	- 1	- 1

Below this row we write how the p-orbital ϕ_1 transforms: Identity E leaves it unchanged, so ϕ_1 remains the

same. When we apply C_2 rotation, ϕ_1 is in the place of ϕ_3 but inverted - therefore we say ϕ_1 becomes "- ϕ_3" (see the illustration above).
Under mirror operation, ϕ_1 comes into the place of ϕ_3, so we write in the column under σ_{xz} "ϕ_3".

Finally a mirror operation in the molecule plane inverts ϕ_1, so it becomes "-ϕ_1"

C_{2v}	E	C_2	σ_{xz}	σ_{yz}
A_2	1	1	- 1	- 1
ϕ_1 ->	ϕ_1	- ϕ_3	ϕ_3	- ϕ_1

"Projections" of ϕ_1

Step 3:
We multiply our "projected orbitals" with the characters of the symmetry we are interested in, in this case A_2:

C_{2v}	E	C_2	σ_{xz}	σ_{yz}	
A_2	1	1	- 1	- 1	
ϕ_1 ->	ϕ_1	- ϕ_3	ϕ_3	- ϕ_1	
$A_2 \times \phi_1$	ϕ_1	- ϕ_3	- ϕ_3	- ϕ_1	$= 2\,(\phi_1 - \phi_3)$ ①

We repeat these steps for ϕ_2:

C_{2v}	E	C_2	σ_{xz}	σ_{yz}	
A_2	1	1	- 1	- 1	
ϕ_2 ->	ϕ_2	- ϕ_2	ϕ_2	- ϕ_2	
$A_2 \times \phi_2$	ϕ_2	- ϕ_2	- ϕ_2	ϕ_2	$= 0$ (no contribution)

We do the same for ϕ_1 and ϕ_2 under B_1:

C_{2v}	E	C_2	σ_{xz}	σ_{yz}	
B_1	1	-1	1	-1	
ϕ_1 ->	ϕ_1	- ϕ_3	ϕ_3	- ϕ_1	
$B_1 \times \phi_1$	ϕ_1	ϕ_3	ϕ_3	ϕ_1	$= 2\,(\phi_1 + \phi_3)$ ②
ϕ_2 ->	ϕ_2	- ϕ_2	ϕ_2	- ϕ_2	
$B_1 \times \phi_2$	ϕ_2	ϕ_2	ϕ_2	ϕ_2	$= 4\,\phi_2$ ③

Finally we get three combinations of $2p_z$ orbitals with the symmetry A_2 (①)and B_1 (② and ③).

Now there is a catch: two combinations of the same symmetry will mix with each other, in this case the two B_1 orbitals !

That means we have to combine them in two ways: adding and subtracting them from each other.

The A_2 symmetry orbital we can call $\Phi_1 = N_1 (\phi_1 - \phi_3)$

(with N_1 as normalization factor, which we do not calculate in this tutorial and is not necessary for our considerations here)

One combination of the two B_1 orbitals is ② + ③: $1/2 (\phi_1 + \phi_3) + \phi_2$

The other combination ② - ③: $1/2 (\phi_1 + \phi_3) - \phi_2$

what gives: $\Phi_2 = N_2 (1/2\phi_1 + \phi_2 + 1/2\phi_3)$

and: $\Phi_3 = N_3 (1/2\phi_1 - \phi_2 + 1/2\phi_3)$

With this information we can draw these three MOs where the π-electrons live:

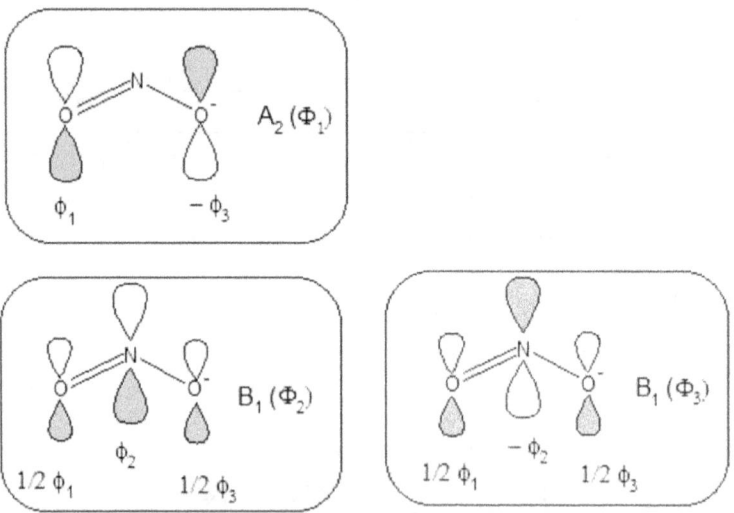

(If we need to find the energies of these orbitals, we would have to calculate the normalization factors N and use matrix algebra. In this tutorial it is enough to know the order of energies, not the absolute energies itself.)

We can use a simple shortcut to find the shape and order of energies for these three π-orbitals. We look on the orbitals from the top:

The MO with all orbitals in phase has the lowest energy and represents bonding combinations:

The next MO is the one that has one node-plane (change of phases in the orbitals) - this MO is non-bonding:

And finally we can draw two node planes - this is the MO with the highest energy and is anti-bonding:

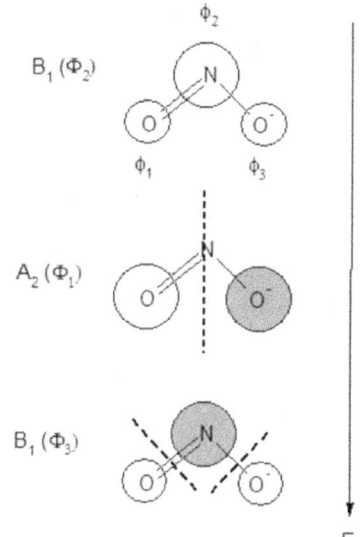

Rules of thumb:

- the sum of the sizes on each atom sort is the same - therefore the two p_z orbitals on O have only half the size than the p_z orbital on N

- a molecule with n MOs can have 0 … (n-1) node planes, in our case 0, 1, 2

- the more nodes, the higher is the energy of the MO - when two MOs have the same number of nodes, their energy is the same (degenerated)

Use the above principles to find out the symmetry, shape and energy order of the π-orbitals in the carbonate anion.

18

Example 2: π electrons of 1,3- butadiene

From the Lewis formula we would expect to see
two distinct double bonds connected by a σ bond.
Therefore the molecule should show a free rotation
around C_2-C_3.

But experimentally there is a significant barrier for rotation. The bond
length is also shorter than a "normal" C-C bond.

We can explain these finding when we analyze the four π electrons,
which are located in the $2p_z$ AOs of the four carbons.

Looking from above as before,
we can identify a totally
symmetric combination.

The next orbital combination
has to use one node to divide
the molecule in half.

The next two combinations
have two and three nodes:

By reducing the complexity of the bonds to only four π electrons allows
to estimate the shape and the energies the MOs in a relatively simple
way.

This simplification is often justified since the electronic properties of the
molecule are mainly due to these π electrons.

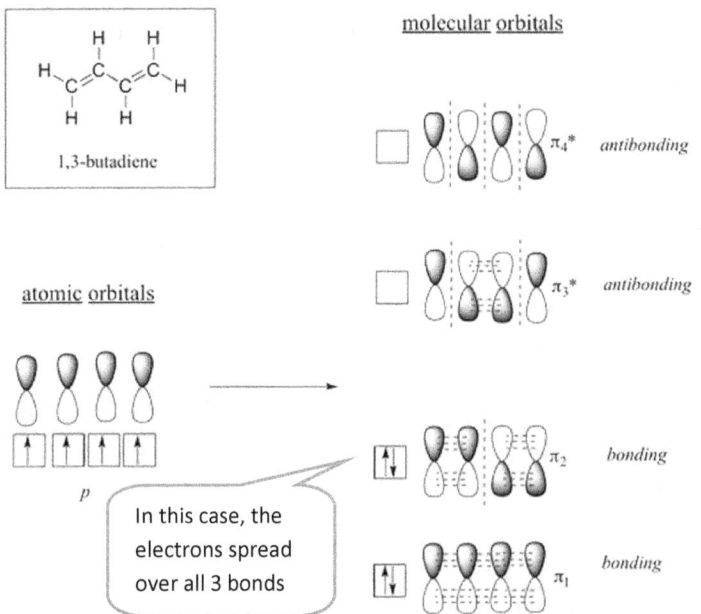

molecular orbitals

π_4^* antibonding

π_3^* antibonding

atomic orbitals

π_2 bonding

p

In this case, the electrons spread over all 3 bonds

π_1 bonding

1,3-butadiene

(http://chemwiki.ucdavis.edu/Core/Organic_Chemistry/Conjugation/ Conjugated_Dienes)

Determine the point group of the Butadiene molecule.   19

If you find out correctly, you should find the point group C_{2h}.

Using the four p-orbitals as basis, find out their representation Γ_π

C_{2h}	E	C_2	i	σ_h	linear
A_g	1	1	1	1	
B_g	1	-1	1	-1	
A_u	1	1	-1	-1	z
B_u	1	-1	-1	1	x,y
Γ_π	4	0	0	-4	

Then we have to use our reduction formula again to see which symmetries are inside the four p_z-orbitals:

$$\Rightarrow \Gamma_\pi = 2\,A_u + 2B_g$$

$A_g = 1/h\,(1*1*4 + 0 + 0 + 1*1*(-4)) = 0$
$B_g = 1/h\,(1*1*4 + 0 + 0 + 1*(-1)*(-4)) = 1/4 * 8 = 2$
$A_u = 1/h\,(1*1*4 + 0 + 0 + 1*(-1)*(-4)) = 2$
$B_u = 1/h\,(*1*4 + 0 + 0 + 1*1*(-4)) = 0$

That means that two electrons are in a A_u state, and two in B_g.

37

How do these four MOs look like ?- we can use our graphical approach:

(1) the MO with the lowest energy has no nodes in the π orbitals, which have the same symmetry as a p_z orbital - from the character table we see it has the representation **A$_{2u}$**. *(remember: "u" means anti-symmetric for inversion i)*

(2) Adding one node, the MO is antisymmetric towards inversion, therefore the representation has also an index

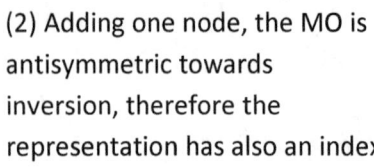
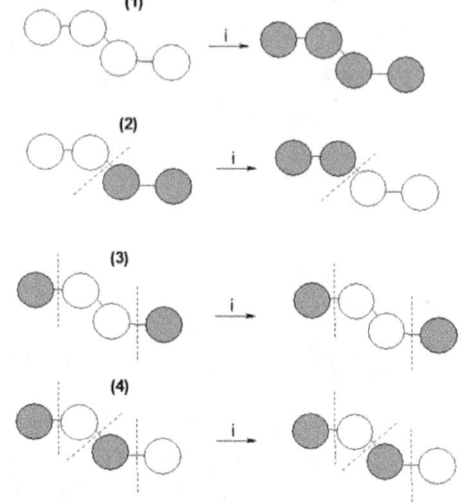

"u" - the representation of the MO with one node must also be **A$_{2u}$**.

Drawing two nodes (3) and three nodes (4) gives MOs symmetric to inversion center, therefore the two energy-highest MOs are represented by **B$_g$**.

Find the representations of the four π MOs of cis-butadiene and their shapes.

20

Transition Metal complexes - ML$_4$

For example we can look at a NiCl$_4$ $^{2-}$ complex:

which obviously has tetrahedral symmetry T$_d$.

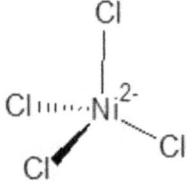

So we have to find out how the 4 ligands transform under these symmetry operations:

T$_d$	E	C$_3$(8)	C$_2$(3)	S$_4$(6)	σ$_d$(6)
Unshifted ligand atoms Γ	4	1	0	0	2

When you calculated right, you should find: $\Gamma = A_1 + T_2$

Inspection of the character table tells us the symmetry of the metal orbitals, which interact with these four group orbitals of the ligands:

Character table for T_d point group

	E	$8C_3$	$3C_2$	$6S_4$	$6\sigma_d$	linear, rotations	quadratic
A_1	1	1	1	1	1		$x^2+y^2+z^2$
A_2	1	1	1	-1	-1		
E	2	-1	2	0	0		$(2z^2-x^2-y^2, x^2-y^2)$
T_1	3	0	-1	1	-1	(R_x, R_y, R_z)	
T_2	3	0	-1	-1	1	(x, y, z)	(xy, xz, yz)

A_1 -> s orbital
(s not explicitly written but the first row **always** represents s symmetry)

T_2 -> p_x, p_y and p_z
and: d_{xy}, d_{xz} and d_{yz}

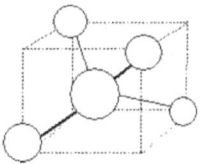

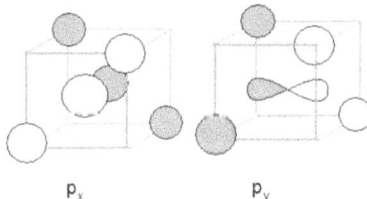

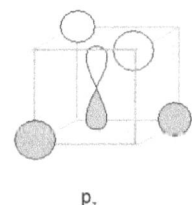

p_x p_y p_z

The three group orbitals for the four Cl ligands T_2 match the symmetry of the metal p_x, p_y and p_z AOs as well as the t_2 d-AOs.

From this information we can draw the MO diagram:

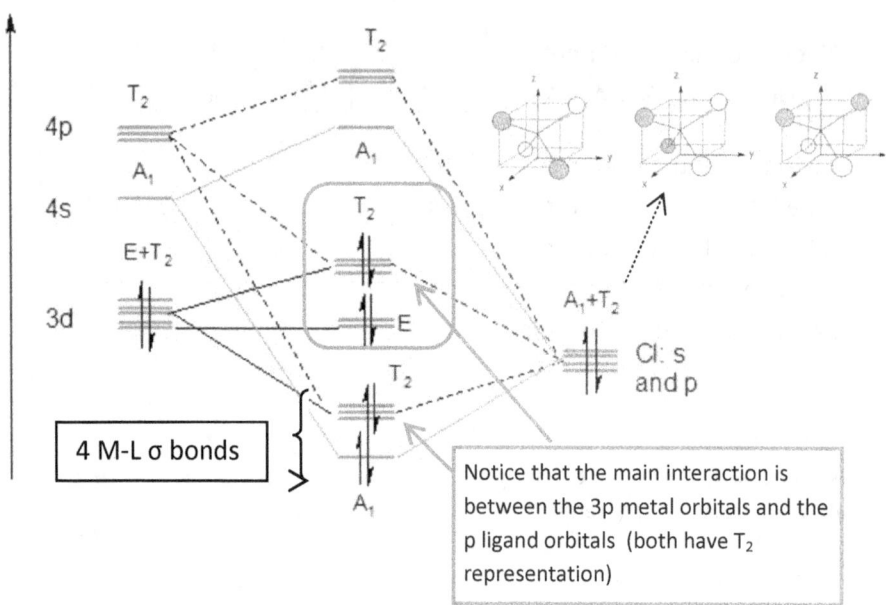

The metal-d electrons are in the E and the T_2 MO-sets. This picture corresponds to the d-orbital "splitting" in Crystal-Field Theory. E is non-bonding and T_2 slightly anti-bonding.

22

Following the same steps as above, construct the MO-diagram for an __octahedral ML_6 complex__ with 6 σ bonded ligands. As a hint - you can construct the ligand group orbitals by introducing one and two node planes:

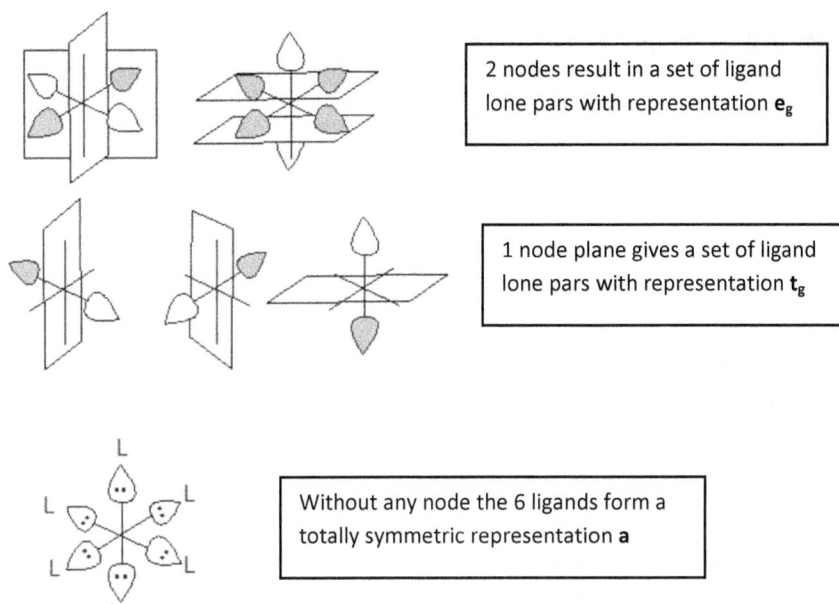

2 nodes result in a set of ligand lone pars with representation e_g

1 node plane gives a set of ligand lone pars with representation t_g

Without any node the 6 ligands form a totally symmetric representation **a**

With this information you should try to match the metal d-, s- and p-orbitals with these group orbitals.

Applications in IR vibrations

Remember that group theory deals with the symmetry of any "thing", not only atoms in a molecule or orbitals. It can be as well be applied to movements.

When the dipole moment of a molecule changes due to vibration, this electric dipole can interact with electromagnetic waves, giving rise to light absorption:

We can treat this oscillating movement with our symmetry operations.

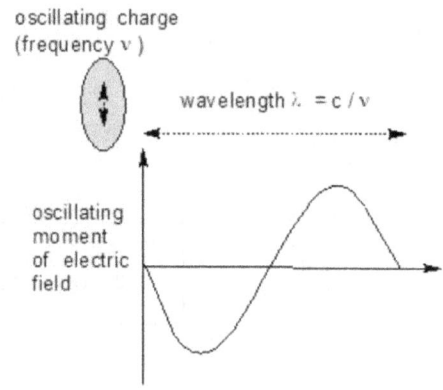

Transition metal carbonyl compounds

An important application is to look about the symmetry of stretching vibrations in transition metal compounds with carbonyl ligands.

These carbonyl ligands give rise to very strong IR signals and which can show us the symmetry of the complex. For example we can distinguish between two planar palladium complexes:

Find out the point group of each of the two isomers

23

42

Then we have to define the coordination system for the molecule:

The z-axis in this case should be perpendicular to the molecule plane. The x- and y-axis are found again by the "right-hand rule".

Next we have to find out how the CO-vibrations behave in that point group. The two arrows in the trans-complex:

D_{2h}	E	$C_2(z)$	$C_2(y)$	$C_2(x)$	i	$\sigma(xy)$	$\sigma(xz)$	$\sigma(yz)$
Γ	2	0	0	2	0	2	2	0

Rotation around the R_3P-Pd-PR_3 line interchanges the two movements

We reduce this representation of the two arrows:

Character table for D_{2h} point group

	E	C_2 (z)	C_2 (y)	C_2 (x)	i	σ (xy)	σ (xz)	σ (yz)	linear, rotations	quadratic
A_g	1	1	1	1	1	1	1	1		x^2, y^2, z^2
B_{1g}	1	1	-1	-1	1	1	-1	-1	R_z	xy
B_{2g}	1	-1	1	-1	1	-1	1	-1	R_y	xz
B_{3g}	1	-1	-1	1	1	-1	-1	1	R_x	yz
A_u	1	1	1	1	-1	-1	-1	-1		
B_{1u}	1	1	-1	-1	-1	-1	1	1	z	
B_{2u}	1	-1	1	-1	-1	1	-1	1	y	
B_{3u}	1	-1	-1	1	-1	1	1	-1	x	

By testing "by hand" we could find out that our combination above is the **sum of A_g and B_{3u}**. Since only B_{3u} has a component in x,y or z direction, we can expect only **one IR vibration peak.**

To see how this vibration looks like, we can either reasoning that there are only two possible stretching motions - symmetric and anti-symmetric:

Since A_g is the totally symmetric representation (and not IR active in this case), B_{3u} must be the second movement.

In cases where it is not that obvious, to find out the stretching motion, we can use the so-called "projection method" (discussed in detail later in this book). In short: we subject one motion vector to the symmetry elements and multiply each transformed vector with the characters of the representation we want to test (in this case B_{3u}):

D_{2h}	E	$C_2(z$	$C_2(y)$	$C_2(x$	i	$\sigma(xy)$	$\sigma(xz)$	$\sigma(yz)$
B_{3u}	1	-1	-1	1	-1	1	1	-1
r_1 becomes:	r_1	r_2	r_1	r_2	r_2	r_1	r_1	r_2
Multiply:	r_1	$-r_2$	$-r_1$	r_2	$-r_2$	r_1	r_1	$-r_2$

Which gives the sum of $2r_1 - 2r_2$ or just $\mathbf{r_1 - r_2}$. That means we combine a positive motion r_1 with a negative motion r_2 leading to:

---

24 *Find the point group of the cis-complex and look up the character table (http://www.webqc.org/symmetry.php)*

Do the same analysis as above: find the representation of the two CO-vibration arrows under this point group and reduce it to get the combination of the irreducible characters. Then you can see how many IR peaks we can expect - if the combination Γ has x, y or z components, these will give rise to an IR light absorption.

When all went well, you should find that we can expect two IR resonances for the cis-complex - therefore the IR spectrum tells us which isomer we have, for example in a reaction solution.

The analysis can be done for all kinds of Carbonyl complexes:

	Coordination number		
	4	5	6
3 Carbonyl ligands			
	OC''''M(CO)(OC)	OC'''' M–CO / OC	CO CO / –M–CO / CO
IR-Peaks	2	1	2
		CO / M–CO / CO	CO / –M–CO / OC
IR-Peaks		3	3
		CO / M–CO / OC	
IR-Peaks		3	
4 Carbonyl ligands			
	CO / OC''''M''''CO / OC	CO / OC'''' M– / OC CO	CO / OC–M–CO / OC
IR-Peaks	1	4	1
		OC'''' M–CO / OC CO	CO / OC–M– / OC CO
IR-Peaks		3	4
5 Carbonyl ligands			
		CO / OC''''M–CO / OC CO	CO CO / OC–M– / OC CO
IR-Peaks		2	3
6 Carbonyl ligands			
			CO CO / OC–M–CO / OC CO
IR-Peaks			1

45

IR vibrations in a small molecule

The general rule applies: the number of vibrations in a non-linear molecule is 3N - 6 and in a linear molecule 3N - 5.

(N = number of atoms)

The reason is that a non-linear molecule like water has 3*3 = 9 possible atom movements. But 3 of these would lead to a rotation of the molecule (around each one of the 3 coordination axis) and 3 of them to a translation of the molecule (in x, y and z direction).

In a linear molecule, we would get one less rotation, the one around the molecular axis itself.

Example 1: water molecule

H_2O is a non-linear molecule with 3*3 - 6 = 3 possible vibration modes.

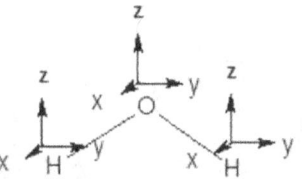

- To find the vibration modes, we first find out the number of unshifted atoms under the symmetry operations of the point group of the molecule (C_{2v})

C_{2v}	E	C_2	σ (xz)	σ (yz)
# unshifted atoms	3	1	3	1

- Use the following information to find the behaviour of the x/y/z coordinate system for each operation:

The characters of the three Cartesian axes under:
- E operations: 3
- C_n operation: $1 + 2\cos(360/n)$
- S_n operation: $-1 + 2\cos(360/n)$
- σ operation: 1
- i operation: - 3

The representation of the atom displacements results from multiplying the two rows.

C_{2v}	E	C_2	σ (xz)	σ (yz)
# unshifted atoms	3	1	3	1
xyz coordinates	3	-1	1	1
Γ_{3N}	9	-1	3	1

(we would get the same result if we considered the symmetry operations on all the 9 coordination vectors in the picture above)

- Reduce the representation Γ_{3N} :
  $n(A_1) = 1/4 [1*1*9) + (1*1*-1) + (1*1*3) + (1*1*1)] =3$
  $n(A_2) = 1 \quad n(B_1) = 2 \quad n(B_2) = 3$

- Examine the character table to find out which of these are translations and rotations:

Character table for C_{2v} point group

	E	C_2 (z)	σ_v(xz)	σ_v(yz)	linear, rotations	quadratic
A_1	1	1	1	1	z	x^2, y^2, z^2
A_2	1	1	-1	-1	R_z	xy
B_1	1	-1	1	-1	x, R_y	xz
B_2	1	-1	-1	1	y, R_x	yz

Translations: $\Gamma_T = A_1 + B_1 + B_2$
Rotations: $\Gamma_R = A_2 + B_1 + B_2$

We subtract these molecule movements from Γ_{3N} :
$\Gamma_{vib} = \Gamma_{3N} - \Gamma_T - \Gamma_R =$
$(3 A_1 + A_2 + 2 B_1 + 3 B_2) -$
$(A_1 + B_1 + B_2) - (A_2 + B_1 + B_2) =$
$2 A_1 + B_2$

- Since A_1 and B_2 contain a linear movement (z and y), all these 3 vibrations give rise to IR absorptions.

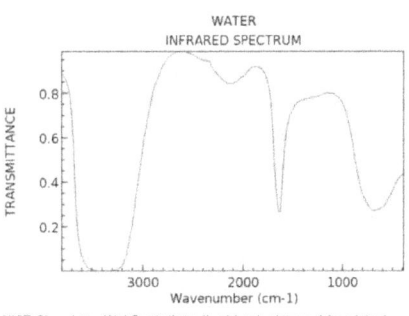

WATER
INFRARED SPECTRUM

NIST Chemistry WebBook (http://webbook.nist.gov/chemistry)

Example 2: NH_3 molecule

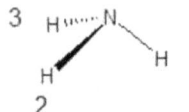

3 ... H ... N

H

H

2

1

- Find the point group and the character table (for example from: http:// webqc.com)

- Find the number of unshifted atoms under each symmetry operation

- Find the characters for the xyz coordinates for each symmetry operation

- Find the reducible representation for the 3N possible motions

- Reduce this representation into irreps

- Find which of these irreps give IR active vibrations

If you did this exercise, then you should find that the representation of all atom displacements is: $\Gamma_{3N} = 3\,A_1 + A_2 + 4\,E$ in the point group C_{3v}.

Vibrations in this atom displacements: $\Gamma_{vib} = \Gamma_{3N} - \Gamma_{trans} - \Gamma_{rot} =$
$= (3\,A_1 + A_2 + 4\,E) - (A_1 + E) - (A_2 + E) = \mathbf{2\,A_1 + 2\,E}$

which corresponds to $2 + 2*2 = 6$ vibrations, that fits with 3N-6 for NH_3.

Finally the character table tells us which of these 6 vibrations result in a peak in the IR spectrum: only these that have a linear component (x, y or z), in this case that is true for all (A_1 -> z and E -> x,y).

But remember that "E" means double degenerate, so one "E" would give only one peak in the spectrum.

Since NH_3 is a gas under normal conditions, the spectrum shows a lot of overlaid rotational peaks.

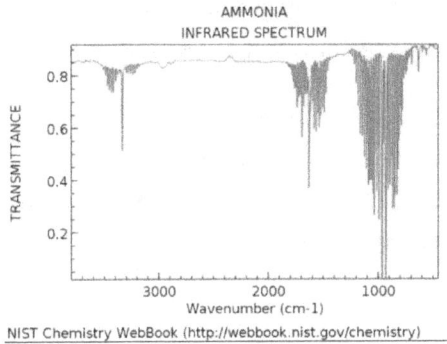

AMMONIA
INFRARED SPECTRUM

NIST Chemistry WebBook (http://webbook.nist.gov/chemistry)

Compare this with the IR
spectrum of solid ammonia:
(*http://web.clark.edu/ggrey/chemst
ruct/IR/
infrared_spectroscopy.htm*)

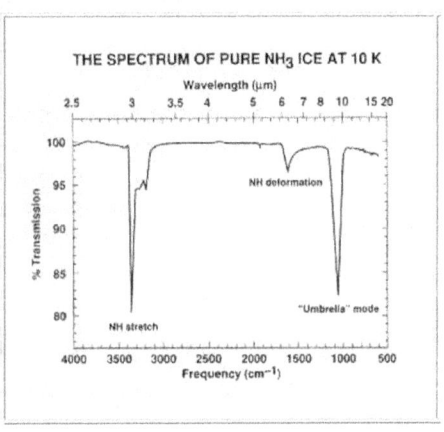

THE SPECTRUM OF PURE NH_3 ICE AT 10 K

26

Exercise: *Find the normal modes of vibration for a PCl_5 molecule*

(hint: consider that there are two different kinds of P-Cl bonds -
equatorial and axial)

Use of internal Coordinates

The two examples above for a water and ammonia molecule show how to find the so-called normal modes of vibrations for a small molecule.

Nevertheless we do not know which of these vibrations are stretching and which are bending movements and how they look like. For this purpose we can use so-called internal coordinates instead of the cartesian coordinates of each atom. Internal coordinates are bonds and angles inside a molecule.

Take the water molecule as example - instead of looking to the 9 cartesian coordinates for each atom, we consider only the two bonds and one bond angle as internal coordinates:

Consider how the two bonds transform under the symmetry operations of the point group C_{2v}:

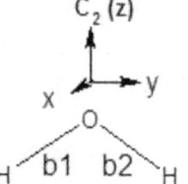

C_{2v}	E	C_2	σ (xz)	σ (yz)
# unshifted bonds	2	0	0	2

This representation includes the stretching vibrations of the molecule, and by reducing it, we find that: $\Gamma_{stretch} = A_1 + B_2$

Since all vibrations are: $\Gamma_{vib} = 2 A_1 + B_2$ (see above), the bending vibration must be: $\Gamma_{vib} = \Gamma_{stretch} + \Gamma_{bend}$

$\Rightarrow \Gamma_{bend} = (2 A_1 + B_2) - (A_1 + B_2) = A_1$

Example: NR$_3$ molecule with point group C$_{3v}$:

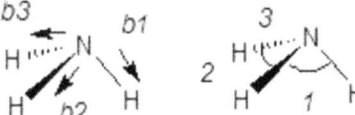

We have to look at 3 stretching and 3 bending motions.

The stretching arrows under the C$_{3v}$ symmetry:

C$_{3v}$	E	2C$_3$	3σ$_v$
Γ$_{N-H}$	3	0	1

Reducing Γ$_{N-H}$ results in

$$\Gamma_{stretch} = A_1 + E$$

Since Γ$_{vib}$ = **2 A$_1$ + 2 E** (see above), the bending vibrations are also:

Γ_{bend} = **A$_1$ + E**

Exercise: *Find the stretching and bending vibrations in PCl$_5$*

$\boxed{27}$

Projection Operations

So far we can find out the vibration modes of small molecules and can distinguish between stretching and bending modes using internal coordinates.

The last step we should follow is to get the actual movement of atoms assigned to these vibrations. To achieve this, we use the so-called method of "projection operators". The meaning is quite simple: we "project" an internal coordinate onto the symmetry operations of the point group. In other words - we check the behaviour of an internal coordinate (like a bond) under the symmetry operations.

Let's look again at the water molecule:
How does the bond b1 transform in the character table of C$_{2v}$:

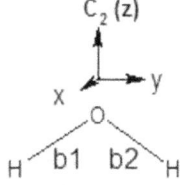

C$_{2v}$	E	C$_2$	σ (xz)	σ (yz)
b1 becomes:	b1	b2	b2	b1

51

We know the two stretching modes are A_1 and B_2, so we have to multiply these characters with the transformed bond:

C_{2v}	E	C_2	σ (xz)	σ (yz)
b1 becomes:	b1	b2	b2	b1
A_1	1	1	1	1

=> the A_1 stretching vibration mode is: 2 b_1 + 2 b_2

That does this mean ? This describes the stretching motion of the two bonds and it says that both bonds become longer and shorter at the same time - therefore it is a symmetric stretching:

The same procedure for the second stretching vibration B_2 gives:

C_{2v}	E	C_2	σ (xz)	σ (yz)
b1 becomes:	b1	b2	b2	b1
B_2	1	-1	-1	1

Which combined together gives: 2 b1 - 2 b2 or normalized:
$1/\sqrt{2}$ (b$_1$ - b$_2$)

This describes the anti-symmetric stretching movement, one bond becomes longer and the other shorter:

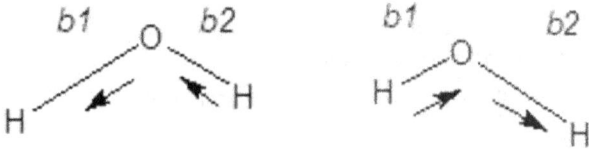

For a slightly more complex molecule as NH_3, we will see that we can use this method also for bond angles:

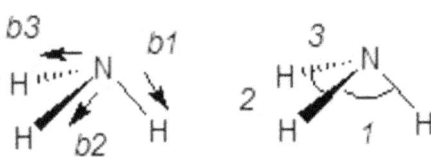

First to find the pattern of **stretching movements**, we use the projection method - find how one of the bonds transforms under each symmetry operation:

C_{3v}	E	C_3	C_3^2	$\sigma_v(1)$	$\sigma_v(2)$	$\sigma_v(3)$
A_1	1	1	1	1	1	1
E	2	-1	-1	0	0	0
b1 becomes:	b1	b2	b3	b1	b2	b3

Totally symmetric vibration:

$P(A_1) = 2 [b(1) + b(2) + b(3)]$
-> $1/\sqrt{3} [b(1) + b(2) + b(3)]$ normalized

and degenerate states:

$P(E) = 2 b(1) - b(2) - b(3)$
-> $1/\sqrt{6} [2 b(1) - b(2) - b(3)]$ normalized

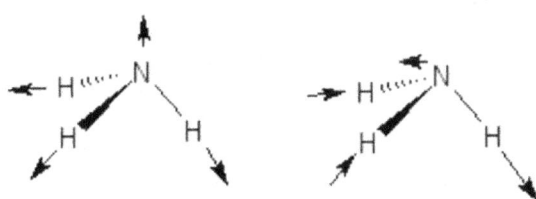

(the problem here is that the movement of the central atom is not included)

In a similar way we can use the 3 bond angles as basis to find the bending movements which again contain A_1 and E.

C_{3v}	E	$2C_3$	$3\sigma_v$
Γ_{H-N-H}	3	0	1

53

To find the bending movements:

C_{3v}	E	C_3	$C_3{}^2$	$\sigma_v(1)$	$\sigma_v(2)$	$\sigma_v(3)$
A_1	1	1	1	1	1	1
E	2	-1	-1	0	0	0
Θ_{H-N-H} (1)	(1)	(2)	(3)	(1)	(2)	(3)

The result of the projection is the same as above: one angle opens and the 2 others close

The general problem with internal coordinates is that in more complex molecules we cannot be sure that we checked all possible stretching and bending coordinates.

28

Find the normal modes of stretching vibrations for $AuBr_4{}^-$
(Find the point group - define internal coordinates - construct the representation of stretching vectors and reduce)

Use the projection method to visualize the stretching modes.

(To evaluate also the bending modes becomes much more difficult because in that case we had to carefully define which bond angles to consider - also some angles are not independent of others)

Electronic Transitions

Electron transitions in a molecule are normally induced by UV or VIS light. In this case light can be absorbed and we can observe a peak in the UV-spectrum, which is determined by the energy that is necessary to induce the transition.

Example: electron transitions in water molecule

Our starting point is again the character table for the point group C_{2v}:

The electric dipole vector, which determines if a light wave is absorbed, transforms as x, y or z.

Character table for C_{2v} point group

	E	C_2 (z)	σ_v(xz)	σ_v(yz)	linear, rotations	quadratic
A_1	1	1	1	1	z	x^2, y^2, z^2
A_2	1	1	-1	-1	R_z	xy
B_1	1	-1	1	-1	x, R_y	xz
B_2	1	-1	-1	1	y, R_x	yz

Therefore a transition between two electronic states absorb light only if their product is one of the components of this dipole vector !

In this example we can examine the transition $A_1 \rightarrow B_2$:

C_{2v}	E	C_2	σ_v(xz	σ_v(yz	
A_1	1	1	1	1	z
B_2	1	-1	-1	1	y
$A_1 \times B_2$	1	-1	-1	1	

We can see that the multiplication of A_1 and B_2 is B_2

(which is true for all multiplications of A_1: is does not change the characters of another irrep)

Since B_2 transforms as the y component, this transition is allowed. It's the transition of an electron from $2p_z$ to $2p_y$ of oxygen.

Example: Benzene molecule

The point group is D_{6h}. Electronic transitions happen most likely as a HOMO-LUMO transfer (Highest Occupied MO and Lowest Unoccupied MO).

So we have first to find the symmetries of these 2 electronic states.

We can apply the graphical method for conjugated π- systems. The lowest energy for the π electrons is the case when they are all in the same phase:

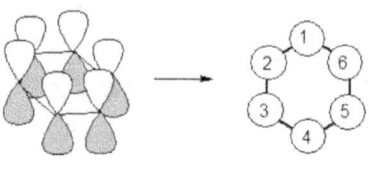

The symmetry of each of the π electrons is A_{2u}: it is the same as the irrep for the z-vector (we are looking at six p_z orbitals).

We can find all possible MOs when we check how all of the $2p_z$ orbitals transform under the operations of the point group D_{6h}:

Character table for D_{6h} point group

	E	$2C_6$	$2C_3$	C_2	$3C'_2$	$3C''_2$	i	$2S_3$	$2S_6$	σ_h	$3\sigma_d$	$3\sigma_v$	Linear, rotations	Quadratic
A_{1g}	1	1	1	1	1	1	1	1	1	1	1	1		x^2+y^2, z^2
A_{2g}	1	1	1	1	-1	-1	1	1	1	1	-1	-1	R_z	
B_{1g}	1	-1	1	-1	1	-1	1	-1	1	-1	1	-1		
B_{2g}	1	-1	1	-1	-1	1	1	-1	1	-1	-1	1		
E_{1g}	2	1	-1	-2	0	0	2	1	-1	-2	0	0	(R_x, R_y)	(xz, yz)
E_{2g}	2	-1	-1	2	0	0	2	-1	-1	2	0	0		(x^2-y^2, xy)
A_{1u}	1	1	1	1	1	1	-1	-1	-1	-1	-1	-1		
A_{2u}	1	1	1	1	-1	-1	-1	-1	-1	-1	1	1	z	
B_{1u}	1	-1	1	-1	1	-1	-1	1	-1	1	-1	1		
B_{2u}	1	-1	1	-1	-1	1	-1	1	-1	1	1	-1		
E_{1u}	2	1	-1	-2	0	0	-2	-1	1	2	0	0	(x, y)	
E_{2u}	2	-1	-1	2	0	0	-2	1	1	-2	0	0		

D_{6h}	E	$2c_6$	$2C$	C_2	$3c'_2$	$3C''_2$	i	$2S_3$	$2S_6$	σ_h	$3\sigma_d$	$3\sigma_v$
E_{1g}	2	1	-1	-2	0	0	2	1	-1	-2	0	0
Γ_π	6	0	0	0	-2	0	0	0	0	-6	2	0

Reducing this representation gives: $\Gamma_\pi = A_{2u} + B_{2g} + E_{1g} + E_{2u}$

We could use the projection method to find how these MOs look like -
test how one of the p$_z$ orbitals transforms under the symmetry
operations and multiply with the characters of the irreps we just found.

For conjugated π-systems however there is a much easier way: the
ground state MO has no node. The next energy level has one node,
which can be drawn in 2 different ways, the same is true for 2 nodes.
Finally 3 nodes gives the highest energy level:

What is the irrep for:

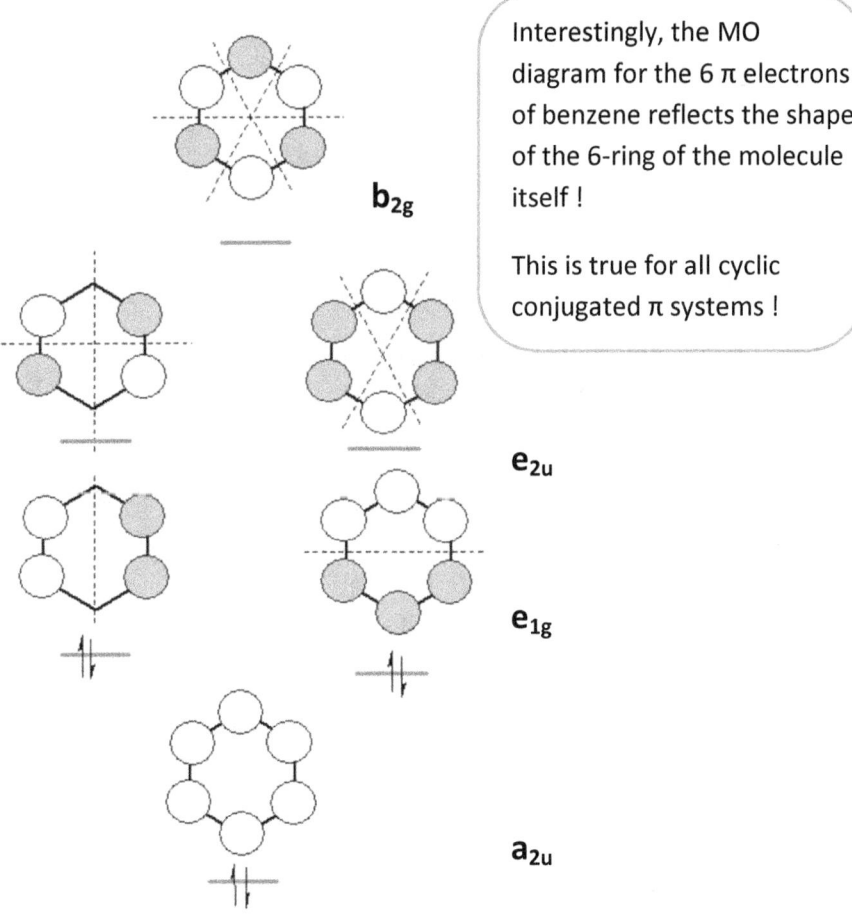

Interestingly, the MO
diagram for the 6 π electrons
of benzene reflects the shape
of the 6-ring of the molecule
itself !

This is true for all cyclic
conjugated π systems !

b$_{2g}$

e$_{2u}$

e$_{1g}$

a$_{2u}$

How can we know which MO belongs to which irrep ?
We should recall in this example that the suffix "g" means symmetric
with respect to inversion, "u" means anti-symmetric.

So we can detect that the two MOs with one node behave symmetric
and the two MOs with two nodes behave anti-symmetric towards
inversion.

Electronic transitions could be:

| $\underline{\quad}$ $\quad$ b_{2g} | $\underline{\quad}$ | $\perp$ | $\underline{\quad}$ | $\perp$ |

$(a_{2u})^2(e_{1g})^4$

Ground State

$e_{1g} \rightarrow e_{2u}$ $\quad$ $e_{1g} \rightarrow b_{2g}$ $\quad$ $a_{2u} \rightarrow e_{2u}$ $\quad$ $a_{2u} \rightarrow b_{2g}$

In spectroscopy, the <u>whole electron distribution</u> can be assigned to an
"electronic state" with a representation on its own.

If all orbitals are occupied as in the ground state, the name of this state
is $^1A_{1g}$: the upper index "1" indicates the spin multiplicity $2S + 1$ ($S =$
total spin of all electrons).

Since S=0 for the ground state, the multiplicity is 1 ("singlet").

The electron density carries the symmetry of the whole molecule,
therefore the state is totally symmetric.

In the case that one electron moves to another energy level, the state is
the product of the representations of the two MOs from where the
electron came and to which it moved - example
$e_{1g} \rightarrow e_{2u}$: the product of these two representations is

D_{6h}	E	$2c_6$	$2C_3$	C_2	$3c'_2$	$3C''_2$	i	$2S_3$	$2S_6$	σ_h	$3\sigma_d$	$3\sigma_v$
E_{1g}	2	1	-1	-2	0	0	2	1	-1	-2	0	0
E_{2u}	2	-1	-1	2	0	0	-2	1	1	-2	0	0
E_{1g} x E_{2u}	4	-1	1	-4	0	0	-4	1	-1	4	0	0

We have to reduce the representation of the product.

Try this out by yourself !

If you calculated right, this product contains the representations **B_{1u}, B_{2u} and E_{1u}** (which represents also the x- and y-axis). The spin S = 0 => multiplicity is 2S + 1 = 1 (the same as the ground state)

Therefore we have to expect 3 possible electronic transitions for the HOMO - LUMO excitation !

$e_{1g} \rightarrow e_{2u}$

This may seem surprising, but we have to consider that it is not the same when the excited electron is in one or the other LUMO's. It makes also a difference which one of the two HOMO electrons is excited.

Selection rules

1. An electronic transition must not change the multiplicity of the system: **ΔS = 0.**
    This condition is called **spin-selection rule**
    (which is fulfilled in our case).

2. **Symmetry selection rule**:
    A transition between two orbitals is orbitally allowed if $\Gamma(n') \rightarrow \Gamma(n)$ transforms the same as one of the components of the dipole operator, i.e., like x, y or z. If the direct product $\Gamma(n') \rightarrow \Gamma(n)$ transforms as x, the electronic transition is said to be "x-polarized" etc.

In our benzene example, only the E_{1u} state represents also the x- and y-vector, but B_{1u} and B_{2u} don't, we can expect that only one transition is symmetry allowed: $^1A_{1g}$ -> $^1E_{1u}$.

3. For centro-symmetric molecules with d-orbitals:
    (typically transition metal complexes)
    the **Laporte rule** says that electronic transitions between "gerade" / "ungerade" states are <u>not allowed</u> (g -> g and u -> u are both forbidden).

Nevertheless, through vibrations a molecule can change its symmetry temporarily and make transitions allowed which should be forbidden according to rule 2 and 3 (but with lower intensity).

The spin-selection rule is much stricter and is usually followed.

Finally the UV spectrum of benzene:
(http://chemwiki.ucdavis.edu/Core/Physical_Chemistry/Spectroscopy/Electronic_Spectroscopy/Electronic_Spectroscopy%3A_Interpretation)

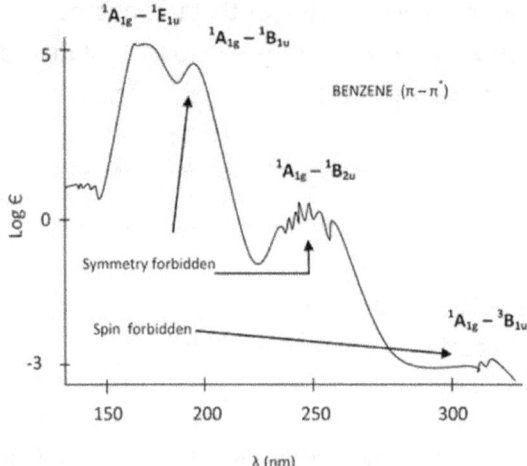

Example cis-butadiene: *Find out the electronic states arising from a HOMO - LUMO transition.*

29

Electronic transitions in transition metal complexes

In transition metal complexes, three different kinds of electronic transitions can be detected:

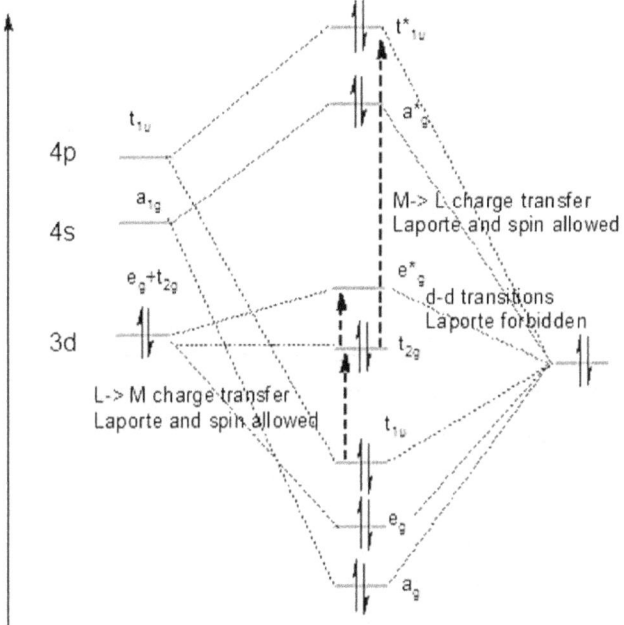

The strongest transitions are the so-called "charge-transfer" where electrons from either the ligands or the metal orbitals move to the metal or ligand MOs.

An obvious feature of many transition metal complexes is their distinct colour. Ni^{2+} complexes for example show strong green colours, Cu^{2+} intense blue colours in aqueous solutions. These visible light absorptions are due to metal-d-d transitions, which are Laporte forbidden (the inversion symmetry of the electronic state does not change, therefore these electron transitions yield only weak signals).

This observation led to a simplified model called "crystal field theory", where the light absorption is explained by electron transitions between metal-d orbitals.

In an octahedral "field" (6 ligands approaching the metal ion):

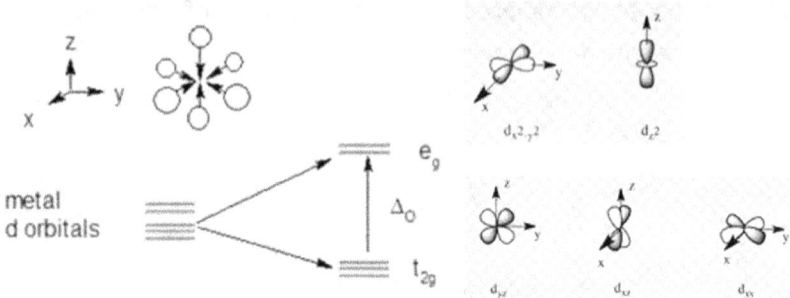

When 6 ligands approach the d-orbitals of the metal ion, then these d-orbitals are repelled by the negative charges on the lone pairs of the ligands. The d_{z2} and d_{x2-y2} point on the x-, y- and z-axis have closest contact to the ligands. Therefore their energy goes up. The other 3 d-orbitals point between the axis and get lower energy.

When we construct the MO diagram for octahedral complexes, then it becomes clear that the t_{2g} orbital set is non-bonding and the e_g set is antibonding. Light is absorbed when an electron is excited from t_{2g} to e_g.

Example: d¹ complexes

The simplest case would be a Ti^{3+} ion with only one electron.

As expected we find one absorption peak in the uv/vis spectrum:

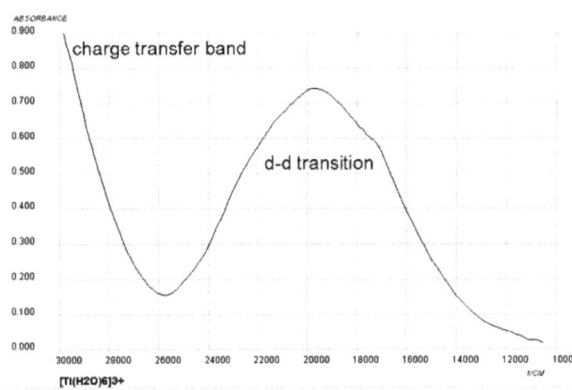

Nevertheless there is a second peak as a shoulder - this is due to Jahn-Teller distortion. This effect is outside the scope of this booklet.

The electron transition looks like this:

The electronic state of the free ion Ti^{3+} is called ²D:

Its multiplicity 2S+1 = 2 ("doublet") and D stands for the orbital quantum number L = 2 (d-orbital).

Under the effect of the 6 ligands, this state ²D splits into e_g and t_{2g} (the five d-orbitals have these kinds of symmetry) and the whole electronic state is called: **²T₂g**

- Index "2" stands for multiplicity = 2S + 1 (in our case 2 * 1/2 +1 = 2)

- "T" stands for 3-fold degenerate energy - in our case the electron can live in each one of the 3 t_{2g} orbitals with the same energy.

- Subindex "2g" describes the symmetry, in our case its the same as the symmetry of the t_{2g} orbitals.

When the electron is excited to the upper level, the whole electronic state is called ^2E_g :

Again the multiplicity is 2 and the electron can now live in two orbitals with the same energy - it is 2-fold degenerated which is expressed by the letter "E".

The splitting into two levels depends on the strength of the ligand field. This relation is quantitatively expressed in a **Tanabe-Sugano diagram**:

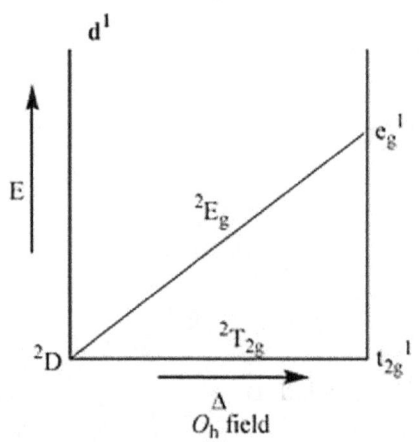

These diagrams can help us to calculate the octahedral splitting energy Δ_O from a uv/vis spectrum - this procedure is beyond the scope of this booklet.

In case of more than one electron, the situation becomes much more complicated because we have to consider the interaction of the d-electrons.

Multi-electron transition metal ions

When we look first at the isolated metal ion, we can determine the term symbols for the ground state:

The letter S, P, D, F etc. results from the angular momentum L (0 = S, 1 = P, 2 = D, 3= F, 4 = G)

d^n	2	1	0	-1	-2	L	S	Ground Term
d^1	↑					2	1/2	^2D
d^2	↑	↑				3	1	^3F
d^3	↑	↑	↑			3	3/2	^4F
d^4	↑	↑	↑	↑		2	2	^5D
d^5	↑	↑	↑	↑	↑	0	5/2	^6S
d^6	↑↓	↑	↑	↑	↑	2	2	^5D
d^7	↑↓	↑↓	↑	↑	↑	3	3/2	^4F
d^8	↑↓	↑↓	↑↓	↑	↑	3	1	^3F
d^9	↑↓	↑↓	↑↓	↑↓	↑	2	1/2	^2D

Under the effect of an octahedral field, these ground terms split differently:

Free Ion Term		Splitting under O_h symmetry
S	→	A_{1g}
P	→	T_{1g}
D	→	$E_g + T_{2g}$
F	→	$A_{2g} + T_{1g} + T_{2g}$
G	→	$A_{1g} + E_g + T_{1g} + T_{2g}$
H	→	$E_g + 2(T_{1g}) + T_{2g}$
I	→	$A_{1g} + A_{2g} + E_g + T_{1g} + 2(T_{2g})$

Starting from a d^2 metal ion, the electronic state ^3F will split in three, not just two levels.

Because there is an interaction between the two electrons, it makes an energetic difference if these electrons are in orbitals of the same plane (xy and x^2-y^2, xz and z^2, yz and z^2) or in perpendicular planes (xy and z^2, xz and x^2-y^2, yz and x^2-y^2) – the electrons have the same spin (Hund's rule) and therefore repel each other. Therefore a configuration with perpendicular planes has slightly less energy.

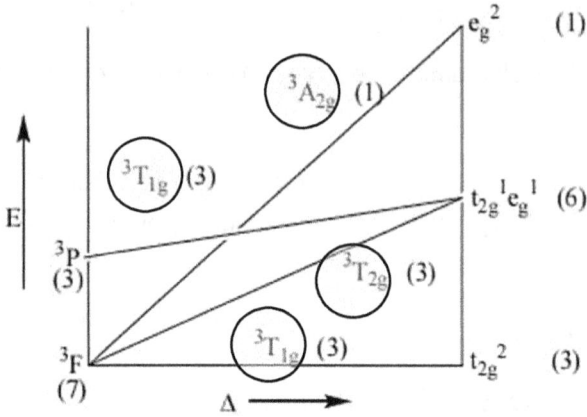

- For the ground state configuration t_{2g}^2, there are 3 different possibilities to place the two electrons, the state symmetry is $^3T_{1g}$

- The excited configuration $e_g^1 t_{2g}^1$ consists of 6 different electronic states ($^3T_{1g}$ and $^3T_{2g}$) - see below.

- The e_g^2 configuration corresponds to an electronic state of $^3A_{2g}$, because there is only one way to fill the electrons (parallel spin).

From the left side, an excited state for the metal ion is ^3P: the sum of the two electron L-values is 1, giving rise to the name "P".

There are two excited states with the configuration $e_g^1 t_{2g}^1$ - they depend on the location of the two electrons which can be in perpendicular orbitals (like d_{xy} and d_{z^2}) (= $^3T_{2g}$) or in orbitals in the same plane (like d_{xy} and d_{x2-y2}) (= $^3T_{1g}$):

66

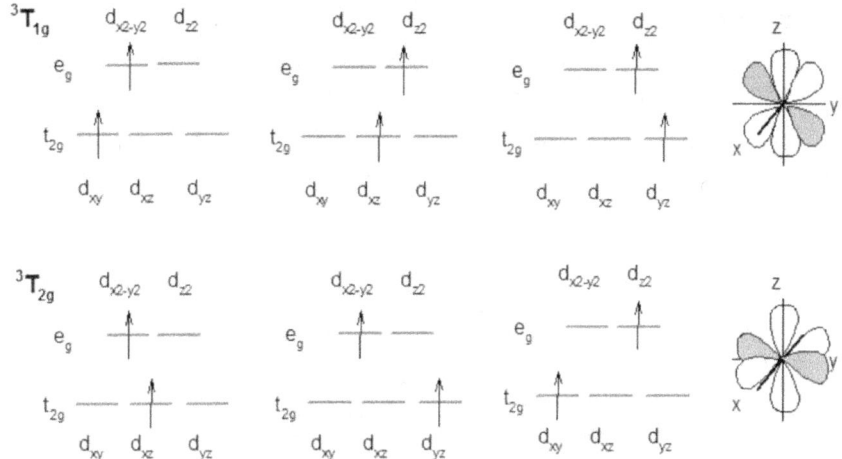

In the (higher energy) $^3T_{1g}$ the orbitals are in the same plane (and have therefore higher repulsion). These orbital combinations are symmetric towards C_2 axis perpendicular the main rotation axis, therefore the index "1".

In the second configuration, the two electrons are in orbitals perpendicular to each other - hence having lower repulsion. These orbital combinations are anti-symmetric towards C_2 axis perpendicular to the main rotation axis, index "2".

If an electron transition is possible or not, we must not look at the electron configurations but on their term symbols - compare a d^1 with a d^2 complex:

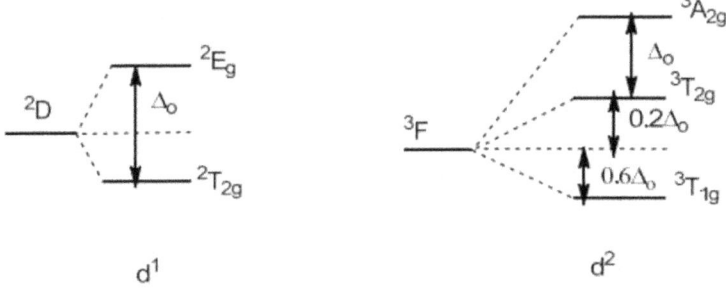

The transition from the ground state $^3T_{1g}$ to $^3T_{2g}$ is spin-allowed, but Laporte-forbidden (both have g symmetry).

The transitions in a V³⁺ complex:

Complex	$^3T_{1g}(F) \rightarrow {}^3T_{2g}$ (cm⁻¹)	$^3T_{1g}(F) \rightarrow {}^3T_{1g}(P)$ (cm⁻¹)	$^3T_{1g}(F) \rightarrow {}^3A_{2g}$ (cm⁻¹)	Δ_o (cm⁻¹)
$[V(H_2O)_6]^{3+}$	17,200	25,000	38,000	21,500

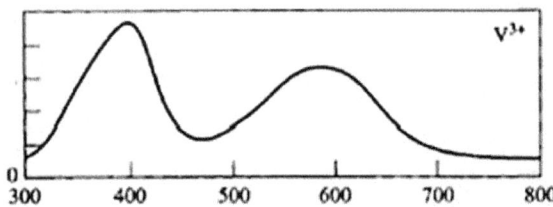

Assign the representation to the following electronic states of a d³ complex:

30

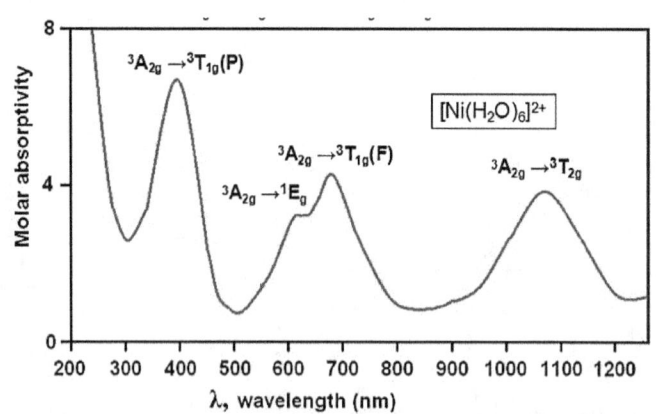

Spectral-Analysis of Transition Metal Complexes

Example for
Ni²⁺ (H₂O)₆:

$^3A_{2g} \rightarrow {}^3T_{1g}(P)$

$[Ni(H_2O)_6]^{2+}$

$^3A_{2g} \rightarrow {}^3T_{1g}(F)$

$^3A_{2g} \rightarrow {}^1E_g$

$^3A_{2g} \rightarrow {}^3T_{2g}$

Molar absorptivity

λ, wavelength (nm)

68

The ground state of this d^8 complex is $^3A_{2g}$.

The two main peaks represent the electron transitions $^3A_{2g} \rightarrow {}^3T_{2g}$:

$^3A_{2g} \rightarrow {}^3T_{2g}$

$= \Delta$
$= 8500$
cm^{-1}

Since this is a transition between "g" electronic states, it is Laporte-forbidden and therefore has only moderate intensity. The energy difference Δ reflects the octahedral splitting energy (but not exactly, see below).

Note that there are in fact two states $^3T_{2g}$ as explained before - so there are two transitions $^3A_{2g} \rightarrow {}^3T_{2g}$.

Another weaker transition at 620 nm is $^3A_{2g} \rightarrow {}^1E_g$:

$^3A_{2g} \rightarrow {}^1E_g$

$= 16100$
cm^{-1}

The peak is even lower because this transition involves a spin-change from triplet (two parallel spins) to singlet (all electrons paired) and it is therefore spin-forbidden in addition to Laporte-forbidden.

This energy change represents the pairing energy of electrons in this level, rather than the Δ_o.

69

In the Ni^{2+} spectrum we can identify <u>three kinds of signals</u>:

1. Laporte- and spin-allowed transitions: charge-transfer signal below 300 nm

2. Laporte-forbidden but spin-allowed transition $^3A_{2g} \rightarrow {}^3T_{2g}$ with moderate intensity

3. Laporte- and spin-forbidden transition $^3A_{2g} \rightarrow {}^1E_g$ with low intensity.

The octahedral splitting energy Δ_o can be exactly determined from the Tanabe-Sugano diagram:

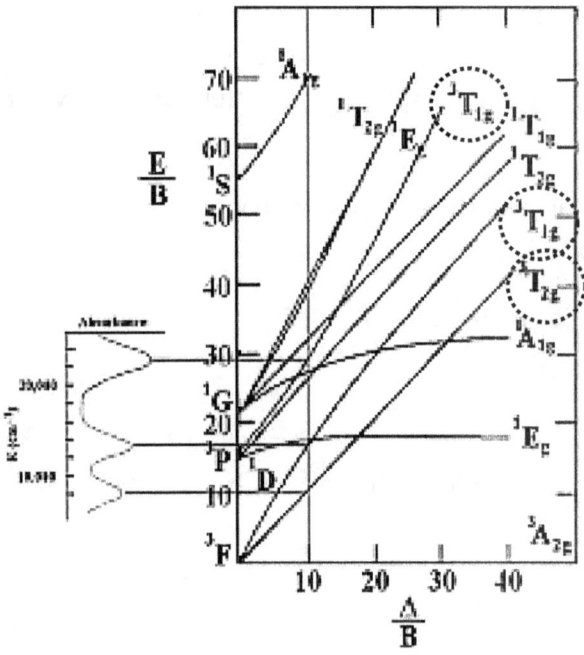

On calculating Δ_o please refer to e.g.
http://wwwchem.uwimona.edu.jm/courses/Tanabe-Sugano/TSHelp.html

A word on tetrahedral complexes (T_d):

The T_d symmetry has no inversion center, therefore the Laporte rule does not apply.

Actually the energetic order of electronic states is reversed compared to the octahedral case

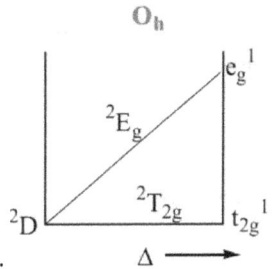 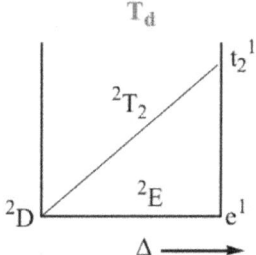

Note that a tetrahedral complex does not have an inversion center, so therefore a d-d transition is not Laporte-forbidden as in an octahedral complex. So these transitions are normally stronger in tetrahedral compounds.

See for example the visible spectrum of Co^{2+} for an octahedral compared to a tetrahedral complex:

The transitions in the tetrahedral complex are not Laporte-forbidden.

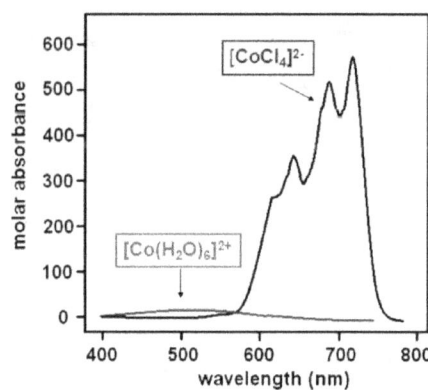

Crystallography

Besides spectroscopic applications, one very important field for symmetry considerations are naturally crystal structures.

In crystallography we do not look at single molecules but on repeating unit cells within a regular solid structure. These unit cells are characterized by distinct symmetry operations which - as before - define under which operation a unit cell remains unaltered.

The unit cell

For example a layer of the cubic NaCl crystal: (2-dimensions)

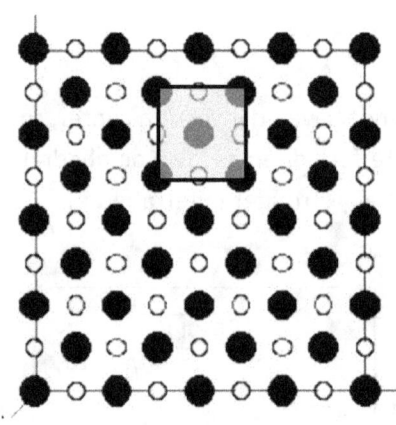

This unit cell contains 4 * 1/2 Na atoms (small) = 2 atoms

and

4 * 1/4 + 1 Cl atoms (big) = 2 atoms

(atoms on the borders are shared by 2 unit cells and count therefore only 1/2, atoms on the edges are shared by 4 unit cells)

Unit cells can have **7 point groups of symmetry**:

C_i C_{2h} D_{2h} D_{3d} D_{4h} D_{6h} O_h

The geometry of a cell group is described by the length of the 3 edges (a,b,c) and the angles between the borders (α,β,γ):

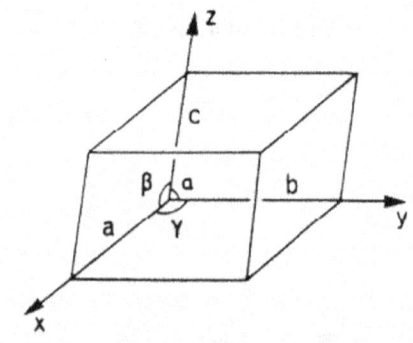

Symmetry in unit cells - Hermann-Mauguin notation

Look at some examples:

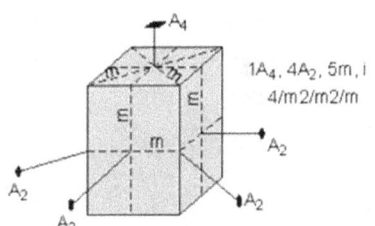

For this unit cell we can identify one C_4 and two distinct rotation axis C_2. There is a mirror plane σ_h perpendicular to each rotation axis.

In crystallography other symbols for symmetry operations are used than in spectroscopy: the Hermann-Mauguin notation for this unit cell would be: 4/m 2/m 2/m $= C_4 + \sigma_h$ and $C_2 + \sigma_h$ and $C_2 + \sigma_h$

Until now we used the so-called "Schoenflies" notation, now we switch to the "Hermann-Mauguin" notation:

Symmetry operation	Schoenflies	Hermann-Mauguin
Identity	E	I
Inversion	i	$-, \overline{1}$
Rotation axis	C_n	n
Mirror plane	σ_h, σ_v	m

In crystallography the symmetry elements are drawn with special symbols:

a rotation axis is drawn with a black polygon with as many edges as the order of rotation.

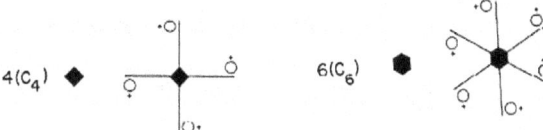

Symmetry element	H-M symbol	graphic symbol
Rotation axes	1	nothing
(n-fold)	2	
	3	
	4	
	5	
	6	
Inversion axes	-1	nothing
	-2 ($\equiv$m)	—
	-3	
	-4	*
	-6	
Mirror plane	m	—

Compare the symmetry elements in the point groups important to crystallography between Schoenflies and Hermann-Mauguin notation:

Hermann-Mauguin nomenclature:

1	2	3	4	6	222	32	422	622	23	432
$\bar{1}$	2/m	$\bar{3}$	4/m	6/m	mmm	$\bar{3}$m	4/mmm	6/mmm	m3	m3m

Schönflies nomenclature:

C_1	C_2	C_3	C_4	C_6	D_2	D_3	D_4	D_6	T	O
C_i	C_{2h}	S_6	C_{4h}	C_{6h}	D_{2h}	D_{3d}	D_{4h}	D_{6h}	T_h	O_h

Practice examples:

Give the Schoenflies and Hermann-Mauguin notations for a point group with

31

(a) a four-fold rotation axis
(b) an inversion axis
(c) a six-fold rotation axis and a mirror plane perpendicular to it

Which symmetry elements can you identify in the point groups:

32

(a) - 1
(b) 3
(c) 6/m
(d) -4/m

Write the Hermann-Mauguin notation for the following unit cell symmetries:
(remember to write the rotation axis before the mirror planes)

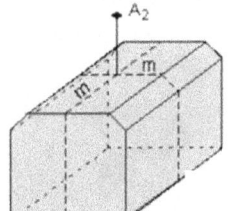

(start with the highest symmetry elements and write the rotation axis before its corresponding mirror plane)

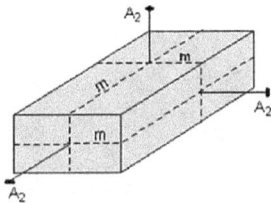

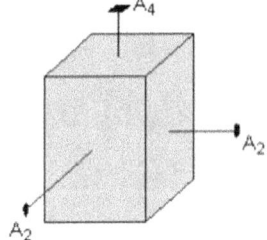

Crystal systems

An overview about the 7 crystal systems with all their points groups:

Crystal System	Herman-Mauguin Point Group	Schoenflies Point Group
Triclinic	1, $\bar{1}$	C_1, C_i
Monoclinic	2, m, 2/m	C_2, C_s, C_{2h}
Orthorhombic	222, mm2, mmm	D_2, C_{2v}, D_{2h}
Trigonal	3, $\bar{3}$, 32,	C_3, S_6, D_3,
	3m, $\bar{3}$m	C_{3v}, D_{3d}
Hexagonal	6, $\bar{6}$, 6/m, 622,	C_6, C_{3h}, C_{4h}, D_6,
	6mm, $\bar{6}$2m, 6mm	C_{6v}, D_{3h}, D_{6h}
Tetragonal	4, $\bar{4}$, 4/m, 422,	C_4, S_4, C_{4h}, D_4,
	4mm, $\bar{4}$2m, 4/mmm	C_{4v}, D_{2d}, D_{4h}
Cubic	23, m3, 432,	T, T_h, O,
	$\bar{4}$32, m $\bar{3}$m	T_d, O_h

The 7 crystal system contain **32 point groups** altogether.

These 7 systems can be put into a hierarchy according to their degree of symmetry (from lowest to highest) :

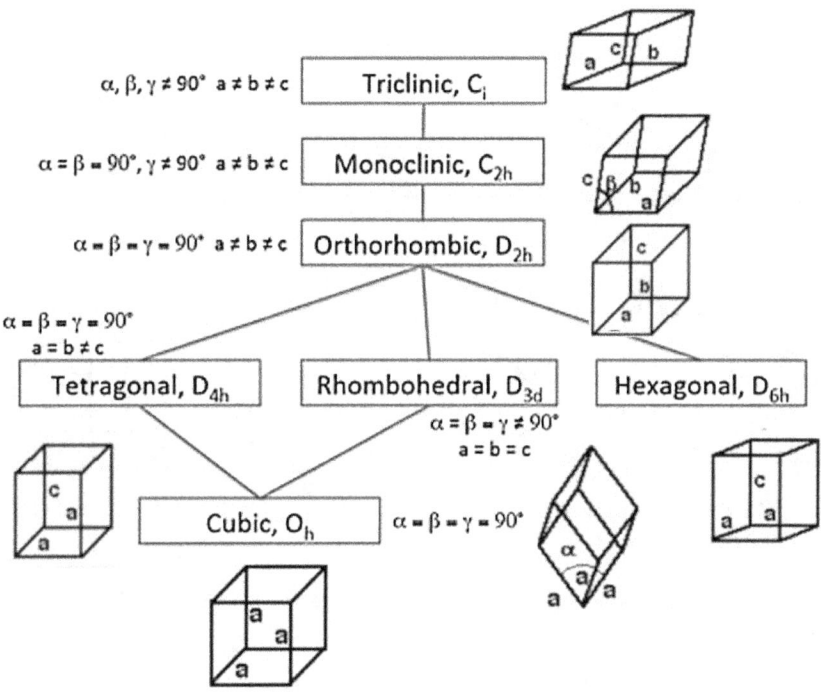

$\alpha, \beta, \gamma \neq 90°$ $a \neq b \neq c$ — Triclinic, C_i

$\alpha = \beta = 90°, \gamma \neq 90°$ $a \neq b \neq c$ — Monoclinic, C_{2h}

$\alpha = \beta = \gamma = 90°$ $a \neq b \neq c$ — Orthorhombic, D_{2h}

$\alpha = \beta = \gamma = 90°$
$a = b \neq c$ — Tetragonal, D_{4h} — Rhombohedral, D_{3d} — Hexagonal, D_{6h}

$\alpha = \beta = \gamma \neq 90°$
$a = b = c$

Cubic, O_h — $\alpha = \beta = \gamma = 90°$

Within these 7 point groups, there can be different placements of atoms (only on the edges or also in the centre of the cell) - this leads to

14 "Bravais" lattices:
(in order of decreasing symmetry)

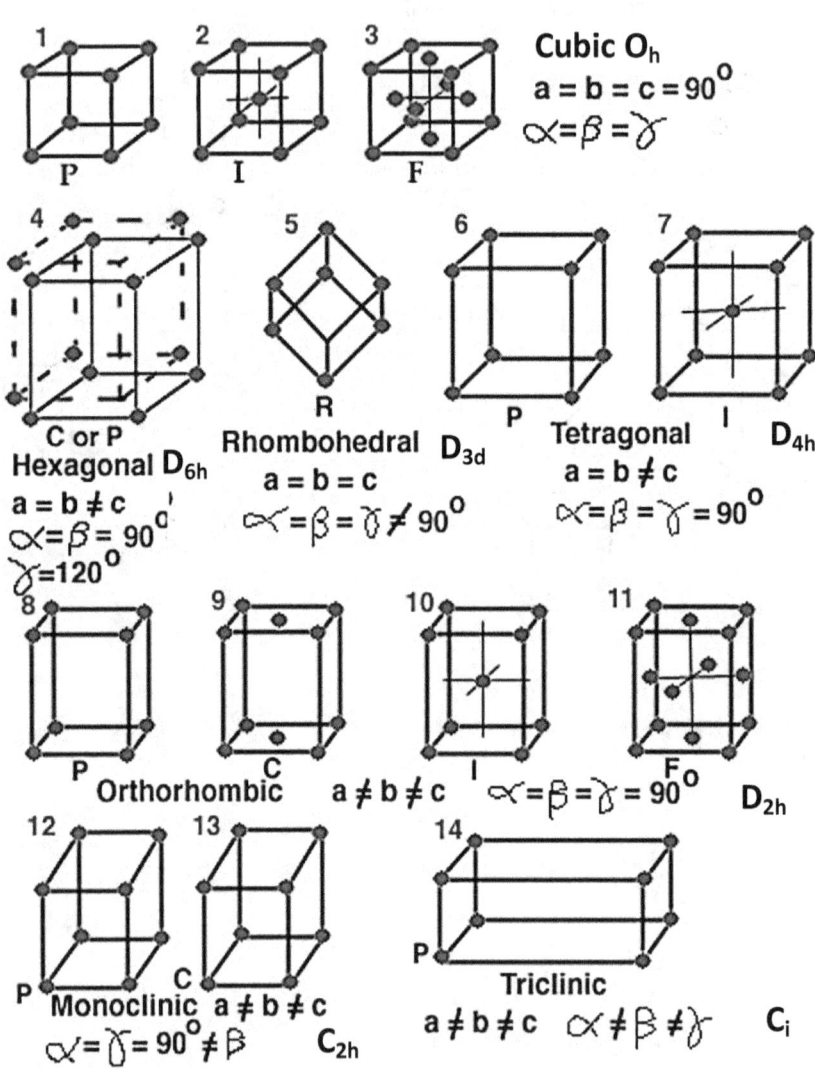

Cubic O_h
$a = b = c = 90°$
$\alpha = \beta = \gamma$

1 P
2 I
3 F

4 C or P
Hexagonal D_{6h}
$a = b \neq c$
$\alpha = \beta = 90°$
$\gamma = 120°$

5 R
Rhombohedral D_{3d}
$a = b = c$
$\alpha = \beta = \gamma \neq 90°$

6 P
7 I
Tetragonal D_{4h}
$a = b \neq c$
$\alpha = \beta = \gamma = 90°$

8 P
9 C
10 I
11 F
Orthorhombic $a \neq b \neq c$ $\alpha = \beta = \gamma = 90°$ D_{2h}

12 P
13 C
Monoclinic $a \neq b \neq c$
$\alpha = \gamma = 90° \neq \beta$ C_{2h}

14 P
Triclinic
$a \neq b \neq c$ $\alpha \neq \beta \neq \gamma$ C_i

Draw all symmetry elements in a monoclinic cell.

34

We distinguish 3 types of unit cells in each case, depending on the position of the atoms:

- Body-centered = "I" (capital i)

- Face-centered = "F"

- Side-centered lattice "C"

- Primitive = "P" (when only the edges of the unit cell carry atoms)

The 14 Bravais lattice types combined with all possible 32 point groups leads to a total of 230 of "space groups".

Further discussion is beyond the scope of this tutorial.

Example NaCl lattice:

The unit cell is a cube with same lengths a, b and c.

The Cl anions (big balls) occupy the edges and the centre of the faces.
The Na cations (small balls) are in the middle of the edges and in the center.

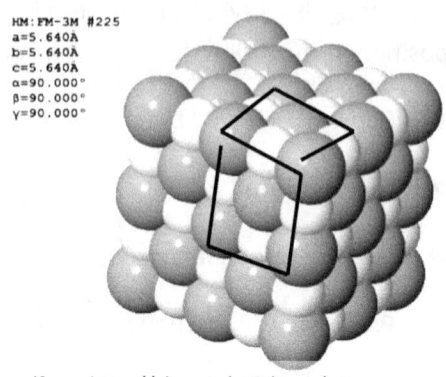

HM: FM-3M #225
a=5.640Å
b=5.640Å
c=5.640Å
α=90.000°
β=90.000°
γ=90.000°

(from http://chemtube3d.com)

Notice that the Na ions form octaheders, as well as the Cl anions !

Lattice type

C	Side face centred
	All face centred
I	Body centred
R	Rhombohedral
P	Primitive

Crystal class

a	triclinic
m	monoclinic
o	orthorhombic
t	tetragonal
h	hexagonal and rhombohedral
c	cubic

Which lattice type and crystal class can you assign to the following

35 unit cells :

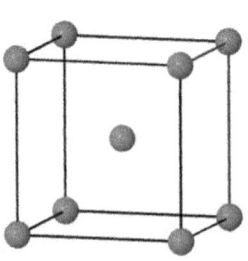

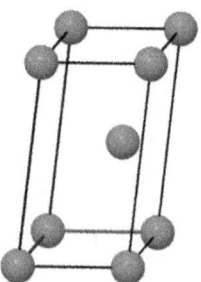

Final words

Finally you worked your way through the main topics in group theory for chemists !

The main goal of this course was to take away the frustration and fear when "normal" chemists are confronted with the strange symmetry symbols like $^2T_{1g}$ etc. Please do not confuse the low capital "t" (which indicates an electronic configuration) and the capital "T" (indicating an electronic state).

Unless you are going to become an expert in spectroscopy, you should now be equipped with a minimum set of knowledge to understand MO-diagrams, IR and UV/VIS spectra as well as crystallographic symbols.

Answers to Questions

Symmetry and Point groups

1 Here are actually 3 rotation axes:

The vertical axis and the one coming out from the paper are c_2 axis.

The axis from left to right is called c_{∞} because there is no definite angle, the ball is symmetric to any rotation angle in this direction.

2 It seems that there is only one rotation axis in z-direction. In fact though there is one axis, there are 2 rotation operations possible, by 120 and by 240 deg (see text in the description)

3 In the ammonia molecule are 3 mirror planes, between each H-atoms

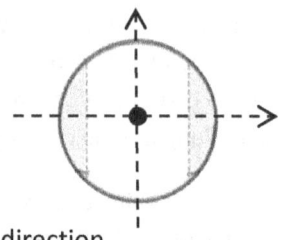

4

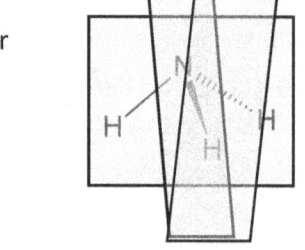

C_2 axis
σ_h and σ_v

Only one σ

C_3 axis
σ_h and 3 σ_v

No free rotation: S_2 axis = i and σ_v

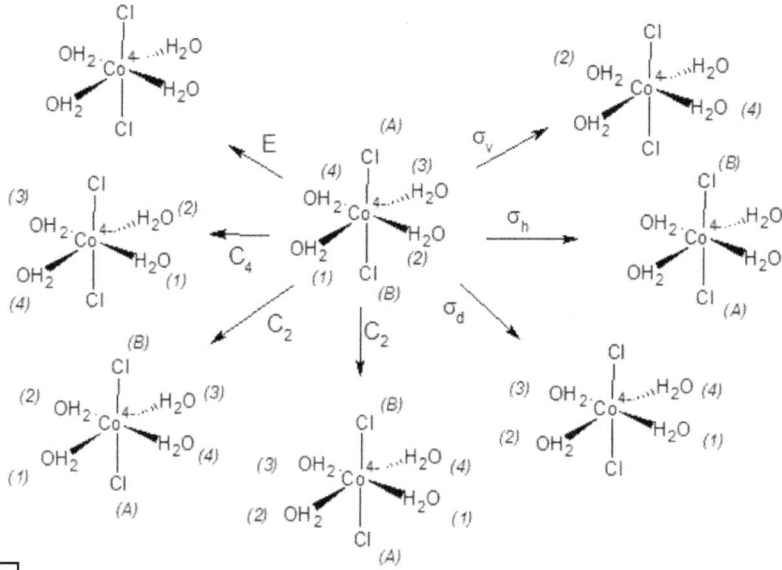

## 6

Point groups:

$SF_5Cl \rightarrow C_{4v}$ $CCl_4 \rightarrow T_d$ $H_2C=CH_2 \rightarrow D_{2h}$ $H_2C=CF_2 \rightarrow C_{2v}$

$SO_4^{2-} \rightarrow D_{4h}$ $SO_3 \rightarrow D_{3h}$ $PCl_5 \rightarrow D_{3h}$ $PCl_3 \rightarrow C_{3v}$

$O=PF_3 \rightarrow C_{3v}$ $(PPh_3)_3RhCl \rightarrow C_{2v}$ mer-$(NH_3)_3(Cl)_3Co \rightarrow C_{3v}$

improper rotation axis through an asymmetric carbon center:
(one of 4 possible ways to draw it)

8 Chiral molecules / asymmetric carbon-centers:

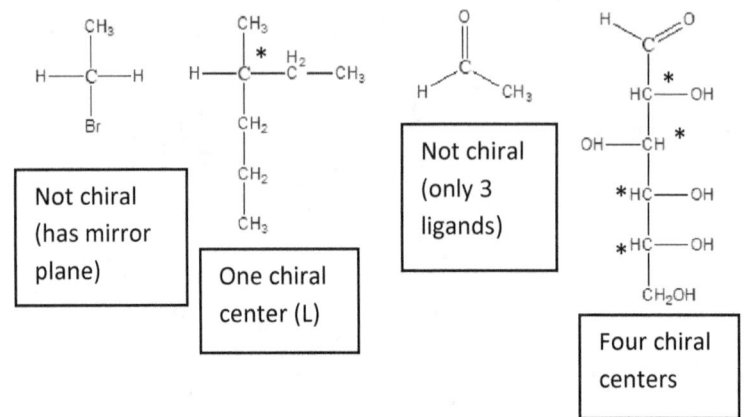

Not chiral (has mirror plane)

One chiral center (L)

Not chiral (only 3 ligands)

Four chiral centers

9 Dipole moments: vector-addition of each dipole moment

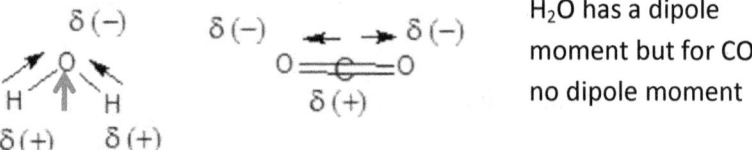

H_2O has a dipole moment but for CO_2 no dipole moment

10 Polarities:

CH_3CH_2-OH (ethanol)	$(CH_3)_2C=O$ (acetone)	CCl_4 (tetra)
3	2	6

CH_3-Cl (chloroform)	C_6H_5-CH_3 (toluene)	$(CH_3)_2$ S=O (DMSO)
4	5	1

The molecules with a C=O and S=O bond have the highest dipole moment due to the difference in electronegativity (EN). The alcohol bond C-OH has a lesser dipole moment than C=O because the hydrogen gives some electron density to the oxygen.

Finally toluene has two different kinds of carbon - sp^2 and sp^3 - which can cause a small dipole moment, compared to tetra which has absolutely no dipole moment.

| 11 | Select solvent for TLC for: |

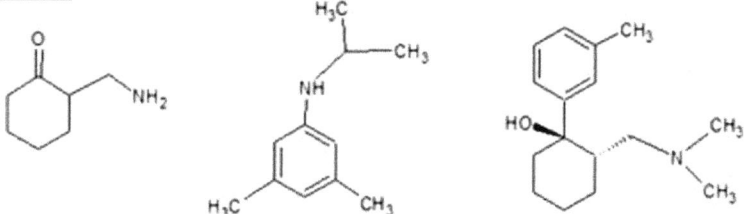

Quite polar: mostly nonpolar: medium polar:

acetone/ethanol   toluene   chloroform

| 12 | Vibrations that cause change in dipole moments: |

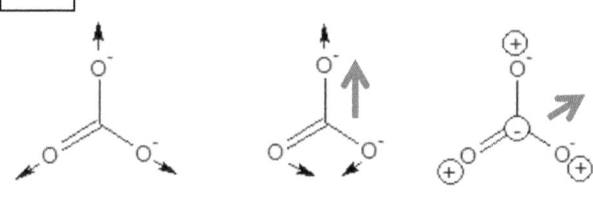

No change   change   change

| 13 | Reducing representations: |

$n(A_2) = 1/4 * [(1*\mathbf{1}*2) + (1*\mathbf{1}*0) + (1*\mathbf{-1}*2) + (1*\mathbf{-1}*0)] = 1/4 * (4 - 4) = 0$

$n(B_1) = 1/4 * [(1*\mathbf{1}*2) + (1*\mathbf{-1}*0) + (1*\mathbf{1}*2) + (1*\mathbf{-1}*0)] = 1/4 * 4 = 1$

$n(B_2) = 1/4 * [(1*\mathbf{1}*2) + (1*\mathbf{-1}*0) + (1*\mathbf{-1}*2) + (1*\mathbf{1}*0)] = 1/4 * (4 - 4) = 0$

(Characters of the irreps in bold)

MO Theory

14	MO diagram of oxygen

The electron filling explains the di-radical character of the oxygen molecule, the bond order is 2 because one bonding electron pair is compensated by one anti-bonding electron pair in π*

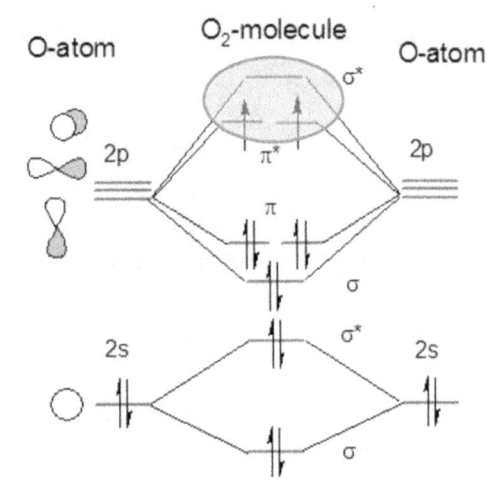

15	MO diagram of NO molecule

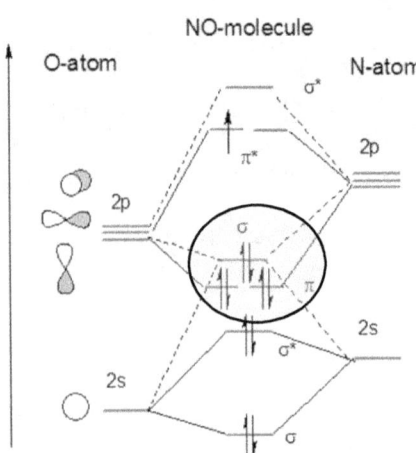

Notice that the AOs of oxygen lie lower in energy than for the nitrogen because of the higher electronegativity of oxygen.

Another important difference to O2 molecule is that mixing of the 2s-AOs with the $2p_z$ AO of oxygen - the energy of the σ MO is higher than for the two π MOs.

## 16

MO Diagrams for Ammonia

Part 1:

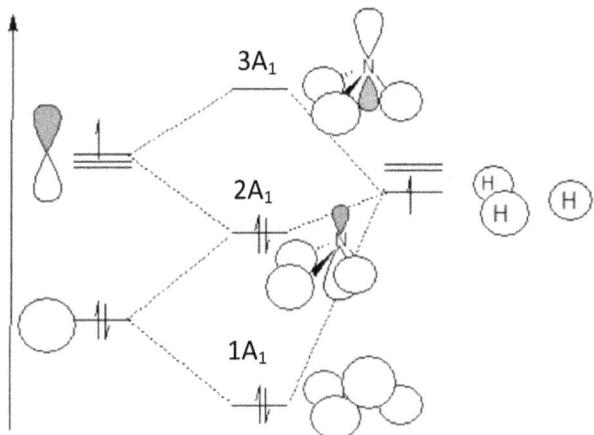

Ammonia MO Diagram Part 2:

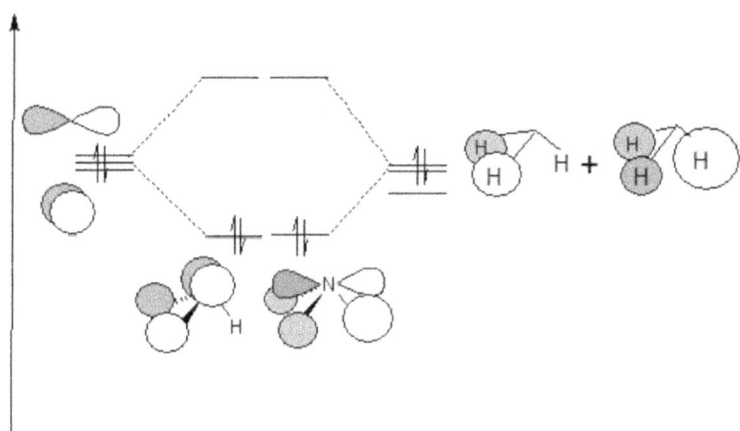

LGOs for PCl$_5$:

(1) Point group is D$_{3h}$:

Character table for D$_{3h}$ point group

	E	2C$_3$	3C'$_2$	σ_h	2S$_3$	3σ_v	linear, rotations	quadratic
A'$_1$	1	1	1	1	1	1		x^2+y^2, z^2
A'$_2$	1	1	-1	1	1	-1	R$_z$	
E'	2	-1	0	2	-1	0	(x, y)	(x^2-y^2, xy)
A''$_1$	1	1	1	-1	-1	-1		
A''$_2$	1	1	-1	-1	-1	1	z	
E''	2	-1	0	-2	1	0	(R$_x$, R$_y$)	(xz, yz)

(2) To construct the group orbitals for the 5 Cl atoms, we can find out how these behave under the symmetry operations here:

D_{3h}	E	$2C_3$	$3C_2$	σ_h	$2S_3$	$3\sigma_v$
Unshifted Cl-bonds Γ_{PCl}	5	2	1	3	0	3

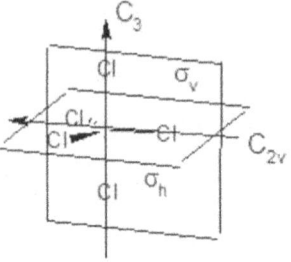

This representation of the Cl atoms can be reduced:

$n(A'1) = 1/12 (5 + 2*2 + 3*1 + 3 + 0 + 3*3) = 2$
$n(A'2) = 1/12 (5 + 2*2 + 3*(-1)*1 + 3 + 0 + 3*(-1)*3) = 0$
$n(E') = 1/12 (2*5 + 2*(-1)*2 + 0 + 2*3 + 0 + 0) = 1$
$n(A''1) = 1/12 (5 + 2*2 + 3*1 + (-1)*3 + 0 + 3*(-1)*3) = 0$
$n(A''2) = 1/12 (5 + 2*2 + 3*(-1)*1 + (-1)*3 + 0 + 3*3) = 1$
$n(E'') = 1/12 (2*5 + 2*(-1)*2 + 0 + (-2)*3 + 0 + 0) = 0$

=> $\Gamma_{PCl} = 2A'_1 + E' + A''_2$

(3) To find these 5 group orbitals by using a projection method would be very tedious. An easier way would be to identify those symmetries in the central Phosphorus atom:

We can assign AOs of the central atom and get two different hybridizations:

$2A'_1$	A''_2	E'	Hybridization
s, d_{z2}	p_z	(p_x, p_y)	sp^3d
s, d_{z2}	p_z	(d_{xy}, d_{x2-y2})	spd^3

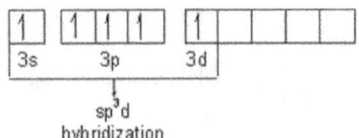

This corresponds to the classical Valence Bond Theory which suggests a **sp³d** hybridization.

To construct the LGOs from reducing Γ_{PCl} would be very tedious. We can construct the 5 group orbitals by the graphical method:

A'_1: fit with s-symmetry

E' fit with p_x and p_y
symmetry

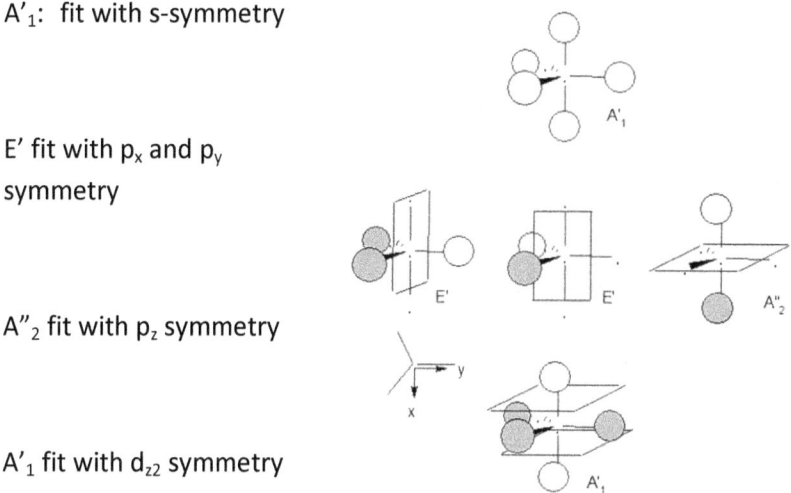

A''_2 fit with p_z symmetry

A'_1 fit with d_{z2} symmetry

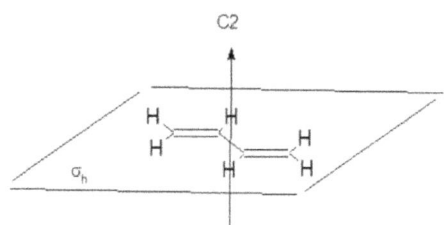

(we can draw **node-planes** into the pictures:
in the upper case, there is no node, for E' and A''_2 one node-plane each
and for A'_1 two node-planes)

| 18 | The point group of trans-butadiene is C_{2h}: |

There is the plane of the molecule itself that forms the horizontal
mirror planl:

19 π-orbitals of the carbonate ion with 6 π electrons in 4 p_z orbitals:

The symmetry is D_{3h}:

Character table for D_{3h} point group

	E	$2C_3$	$3C'_2$	σ_h	$2S_3$	$3\sigma_v$	linear, rotations	quadratic
A'_1	1	1	1	1	1	1		x^2+y^2, z^2
A'_2	1	1	-1	1	1	-1	R_z	
E'	2	-1	0	2	-1	0	(x, y)	(x^2-y^2, xy)
A''_1	1	1	1	-1	-1	-1		
A''_2	1	1	-1	-1	-1	1	z	
E''	2	-1	0	-2	1	0	(R_x, R_y)	(xz, yz)

Now we have to find how the 4 p_z orbitals transform under these symmetry operations:

D_{3h}	E	$2C_3$	$3C_2$	σ_h	$2S_3$	$3\sigma_v$
Γ_p	4	1	-2	-4	-1	2

Reduction of Γ_p :

$n(A'_1) = 1/12\ (4 + 2*1*1 + 3*1*(-2) + 1*1*(-4) + 2*1*(-1) + 3*1*2) = 0$

$n(A'_2) = 1/12\ (4 + 2*1*1 + 3*(-1)*(-2) + 1*1*(-4) + 2*1*(-1) + 3*(-1)*2) = 0$

$n(E') = 1/12\ (2*4 + 2*(-1)*1 + 0 + 1*2*(-4) + 2*(-1)*(-1) + 0) = 0$

$n(A''_1) = 1/12\ (4 + 2*1*1 + 3*1*(-2) + 1*(-1)*(-4) + 2*(-1)*(-1) + 3*(-1)*2) = 0$

$n(A''_2) = 1/12\ (4 + 2*1*1 + 3*(-1)*(-2) + 1*(-1)*(-4) + 2*(-1)*(-1) + 3*1*2) = 2$

$n(E'') = 1/12\ (2*4 + 2*(-1)*1 + 0 + 1*(-2)*(-4) + 2*1*(-1) + 0) = 1$

That means the four p_z orbitals form three combinations $\Gamma_p = 2\ A''_2 + E''$ which are in fact <u>four MOs</u> (since E" represents two MOs with the same energy)

.

To see how these MOs look like, we can use the simple graphical approach and add 0, 1 and 2 node planes to the molecule:

A_2''

To get two equivalent MOs (E"), it is only possible when there is one node plane, which can be drawn vertically or horizontally. For the other two MOs with no and with two node planes, there is only one possiblity.

E"

A_2''

E

| 20 | Point Group of cis-Butadiene: C_{2v} |

Construct the π orbitals of the 4 π electrons:

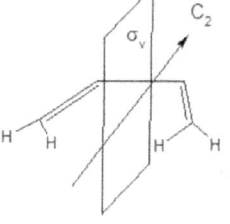

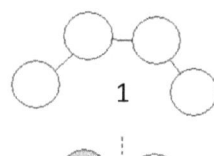

1

As before, we first set all 4 π orbitals in the same phase.

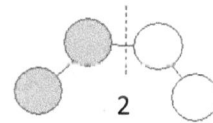

2

Then we split the molecule with one node,

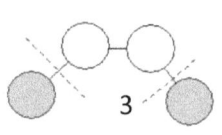

3

and we introduce two nodes,

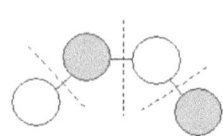

and finally three nodes.

Then we use the character table to find the representations that are included in the 4 π orbitals:

C$_{2v}$	E	C$_2$	σ$_v$(xz)	σ$_v$(yz)	linear
A$_1$	1	1	1	1	z
A$_2$	1	1	-1	-1	
B$_1$	1	-1	1	-1	x
B$_2$	1	-1	-1	1	y
Γ_π	4	0	0	-4	

Γ_π is the same as for trans-butadiene, but the representations are different in this group.

So when we reduce Γ_π we get: Γ_π = 2 A$_1$ + 2 B$_2$

That means that the 4 π electrons live in 2 different symmetry representations. To assign the above MOs with their characters, we can look at a symmetry element that is different between A$_1$ and B$_2$, for example the behaviour towards C$_2$ rotation:

MOs 1 and 3 would be anti-symmetric, therefore they are **b$_2$**,

MOs 2 and 4 are symmetric towards rotation, therefore they are **a$_1$**.

21   Four ligands in NiCl$_4$ $^{2-}$:

T$_d$	E	C$_3$(8)	C$_2$(3)	S$_4$(6)	σ$_d$(6)
A$_1$	1	1	1	1	1
T$_2$	3	0	-1	-1	1
Γ	4	1	0	0	2

n(A$_1$) = 1/24 (1*1*4 + 8*1*1 + 3*1*0 + 6*1*0 + 6*1*2) = 1
n(T$_2$) = 1/24 (1*3*4 + 8*0*1 + 3*-1*0 + 6*-1*0 + 6*1*2) = 1

| 22 | Interaction MO diagram between transition metal s, p and d orbitals with the 6 group orbitals of a two-electron donor ligand: |

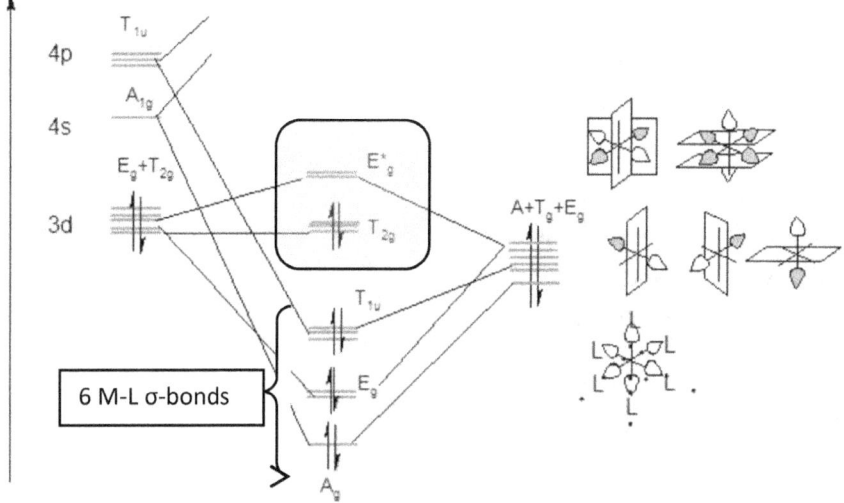

Notice that the LUMO is actually anti-bonding and that the molecule has the best stability if there are 18 valence electrons (thus the famous "18-electron-rule" for transition metal compounds)

IR vibrations

| 23 | Point groups of: |

D$_{2h}$ and C$_{2v}$

24	Cis-Pd complex with point group C_{2v}

C_{2v}	E	C_2	$\sigma\,(xz)$	$\sigma\,(yz)$
Γ_{CO}	2	0	2	0

Reduction leads to: $\Gamma_{CO} = A_1 + B_1$

Both representations transform as a linear vector (z and x), therefore both vibrations are IR active.

Character table for C_{2v} point group

	E	C_2 (z)	$\sigma_v(xz)$	$\sigma_v(yz)$	linear, rotations	quadratic
A_1	1	1	1	1	z	x^2, y^2, z^2
A_2	1	1	-1	-1	R_z	xy
B_1	1	-1	1	-1	x, R_y	xz
B_2	1	-1	-1	1	y, R_x	yz

25	Ammonia molecule stretching vibrations:

Point group C_{3v} :

Character table for C_{3v} point group

	E	$2C_3$ (z)	$3\sigma_v$	linear, rotations	quadratic
A_1	1	1	1	z	x^2+y^2, z^2
A_2	1	1	-1	R_z	
E	2	-1	0	(x, y) (R_x, R_y)	(x^2-y^2, xy) (xz, yz)

C_{3v}	E	C_3	C_3'	$\sigma(xz)$	σ	σ
xyz	3	0	0	1 (*)	0	0
# unshifted atoms	4	1	1	2	2	2
multiply: Γ_{3N}	12	0	0	4	0	0

(*) mirror on the xz plane reverses the y-axis (-1) and leaves the x- and z-axis (2) => character is 2-1 = 1

In total there are 3N = 12 atom movements and 3N-6 = 6 vibrations

Reduction leads to: $\Gamma_{3N} = 3A_1 + A_2 + 4E$ (E counts as 2)

Rotations: $A_2 + E$ (R_z and R_x, R_y) nd Translations: $A_1 + E$ (z and x,y)

=> Vibrations: $\Gamma_{vib} = \Gamma_{3N} - \Gamma_{rot} - \Gamma_{trans} = $ **2 A_1 + 2 E**

Normal modes for PCl$_5$

Point group D$_{3h}$

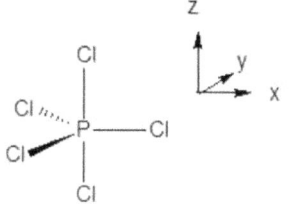

D$_{3h}$	E	2C$_3$	3C$_2$	σ$_h$	2S$_3$	3σ$_v$
xyz coordinates	3	0	-1 (*)	1	-2	1 (**)
# unshifted atoms	6	2	2	4	1	4
multiply Γ$_{3N}$	18	0	-2	4	-2	4

(*) C$_2$ (x) will invert y and z axis but preserve x
(**) σ$_v$(xz) will preserve the x- and z-axis but inverts the y-axis

In total there are 3N = 3*6 = 18 atom movements and 3N - 6 = 12 vibrations:

Reducing Γ$_{3N}$ yields:

Γ$_{3N}$ = 2 A$_1'$ + A$_2'$ + 4 E' + 3 A$_2''$ + 2 E''
Rotations: Γ$_{rot}$ = A$_2'$ + 2 E''
Translations: Γ$_{trans}$ = 2 E' + A$_2''$

Character table for D$_{3h}$ point group

	E	2C$_3$	3C'$_2$	σ$_h$	2S$_3$	3σ$_v$	linear, rotations	quadratic
A'$_1$	1	1	1	1	1	1		x^2+y^2, z^2
A'$_2$	1	1	-1	1	1	-1	R$_z$	
E'	2	-1	0	2	-1	0	(x, y)	(x^2-y^2, xy)
A''$_1$	1	1	1	-1	-1	-1		
A''$_2$	1	1	-1	-1	-1	1	z	
E''	2	-1	0	-2	1	0	(R$_x$, R$_y$)	(xz, yz)

=> Vibrations: Γ$_{vib}$ = Γ$_{3N}$ - Γ$_{rot}$ - Γ$_{trans}$ = **2 A$_1'$ + 3 E' + 2 A$_2''$ + E''**

27 Stretching and bending vibrations in PCl$_5$ using internal coordinates:

D$_{3h}$	E	2C$_3$	3C$_2$	σ_h	2S$_3$	3σ_v
# unshifted bonds	5	2	1	3	0	3

Reduce this representations: $\Gamma_{bonds} = \Gamma_{stretch} = $ **2 A$_1'$ + E' + A$_2$"**
(E' and A$_2$" are IR active because these transform as x,y and z)

From the last exercise we know that $\Gamma_{vib} = $ 2 A$_1'$ + 3 E' + 2 A$_2$" + E"
=> bending modes are $\Gamma_{bend} = \Gamma_{vib} - \Gamma_{stretch} = $ **2 E' + A$_2$" + E"**

28 Vibration modes for the square-planar AuBr$_4^-$

with the point group D$_{4h}$

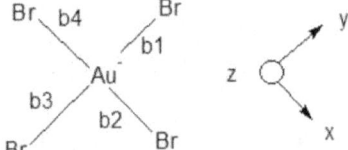

Character table for D$_{4h}$ point group

	E	2C$_4$ (z)	C$_2$	2C'$_2$	2C"$_2$	i	2S$_4$	σ_h	2σ_v	2σ_d	linears, rotations	quadratic
A$_{1g}$	1	1	1	1	1	1	1	1	1	1		x^2+y^2, z^2
A$_{2g}$	1	1	1	-1	-1	1	1	1	-1	-1	R$_z$	
B$_{1g}$	1	-1	1	1	-1	1	-1	1	1	-1		x^2-y^2
B$_{2g}$	1	-1	1	-1	1	1	-1	1	-1	1		xy
E$_g$	2	0	-2	0	0	2	0	-2	0	0	(R$_x$, R$_y$)	(xz, yz)
A$_{1u}$	1	1	1	1	1	-1	-1	-1	-1	-1		
A$_{2u}$	1	1	1	-1	-1	-1	-1	-1	1	1	z	
B$_{1u}$	1	-1	1	1	-1	-1	1	-1	-1	1		
B$_{2u}$	1	-1	1	-1	1	-1	1	-1	1	-1		
E$_u$	2	0	-2	0	0	-2	0	2	0	0	(x, y)	

To examine the stretching vibrations we look at the four Au-Br bonds, how they transform under the operations in D$_{4h}$:

D$_{4h}$	E	2C$_4$	C$_2$	2C$_{2'}$	2C$_2$"	i	2S$_4$	σ_h	2σ_v	2σ_d
# unshifted bonds	4	0	0	2	0	0	0	4	2	0

Reduction yield: $\Gamma_{bonds} = $ A$_{1g}$ + B$_{1g}$ + E$_u$

Apply the projection of one bond onto these symmetries:

D_{4h}	E	C_4^1	C_4^3	C_2	$C_2(x)$	$C_2(y)$	$C_2''(1)$	$C_2''(2)$
b1 ->	(1)	(2)	(4)	(3)	(3)	(1)	(2)	(4)
A_{1g}	1	1	1	1	1	1	1	1
B_{1g}	1	-1	-1	1	1	1	-1	-1
E_u	2	0	0	-2	0	0	0	0
D_{4h}	i	S_4^1	S_4^3	σ_h	σ_v	σ_v'	σ_d	σ_d'
b1 ->	(3)	(2)	(4)	(1)	(1)	(3)	(2)	(4)
A_{1g}	1	1	1	1	1	1	1	1
B_{1g}	1	-1	-1	1	1	1	-1	-1
E_u	-2	0	0	2	0	0	0	0

Under A_{1g} the bond b1 transforms to : 4 b1 + 4 b2 + 4 b3 + 4 b4
and under B_{1g}: 4 b1 - 4 b2 + 4 b3 - 4 b4
and under E_u: 4 b1 - 4 b3

The second mode under E_u can be evaluated when we use bond b2 as a basis - then we get: 4 b2 - 4 b4

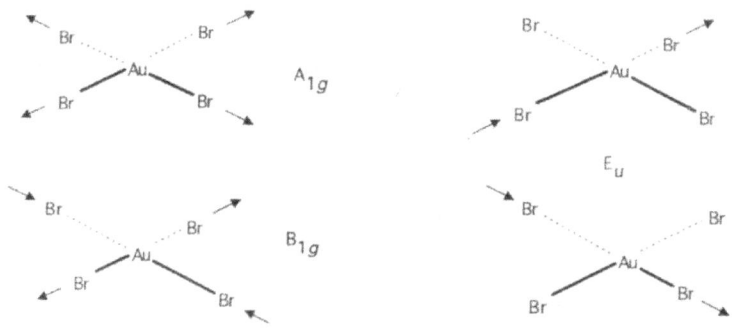

Only the vibration mode E_u is IR active, therefore we expect only one peak in the stretching region of the spectrum.

Electronic transitions

29	cis-Butadiene: possible HOMO-LUMO transitions

From the answer of question 18 we know already that the HOMO as the representation a_1 and the LUMO b_2.

Then we have to form the product of these two representations and check if it contains a linear symmetry:

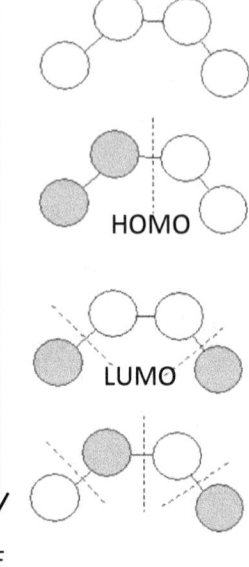

C_{2v}	E	C_2	$\sigma\,(xz)$	$\sigma\,(yz)$
A_1	1	1	1	1
B_2	1	-1	-1	1

Remember: the multiplication of a row with the totally symmetric row would not change it, therefore the product is again **B_2**.

And the irrep B_2 transforms as the y vector, therefore we can expect a HOMOLUMO transition

(don't confuse the <u>electron transition</u> $a_1 \rightarrow b_2$ with the change in <u>electronic state</u> $^1A_1 \rightarrow {}^1B_2$)

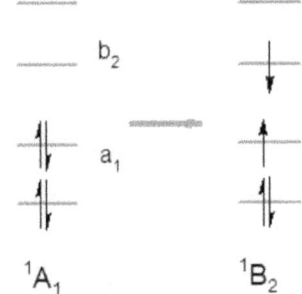

Electronic states of d^3 complexes:

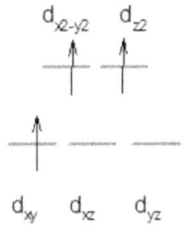

Only one configuration possible => A

Total spin 3/2 => multiplicity 2S + 1 = 4

=> ^{4}A state

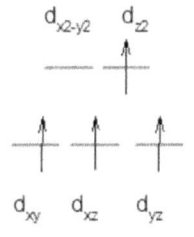

Three possible configurations => T

Total spin 3/2 => multiplicity 2S + 1 = 4

=> ^{4}T state

Two possible configurations => E

Total spin 4/2 => multiplicity 2S + 1 = 5

=> ^{5}E state

Crystallography

| 31 | Point groups for: |

(a) a four-fold rotation axis: C_4 or 4 group
(b) an inversion axis: C_i or -1 group
(c) a six-fold rotation axis and a mirror plane perpendicular to it:
     D_{6h} or 6/m group

| 32 | |

Which symmetry elements can you identify in the point groups:
(a) - 1   inversion center
(b) 3   three-fold rotation axis
(c) 6/m   six-fold rotation axis and perpendicular mirror plane
(d) -4/m S_4 axis and perpendicular mirror plane

| 33 | Hermann-Maugin groups: |

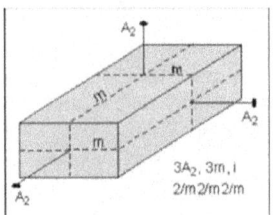

2 m m

(two-fold axis and two mirror planes
containing the axis)

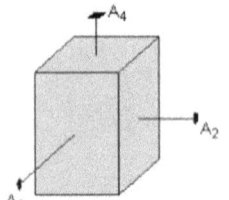

2/m 2/m 2/m

(three two-fold axis and each has a
perpendicular mirror plane)k

422

(four-fold and two two-fold rotation axis)

| 34 | Monoclinic cell: has a two-fold |

100

rotation axis and a perpendicular mirror plane (2 /m):

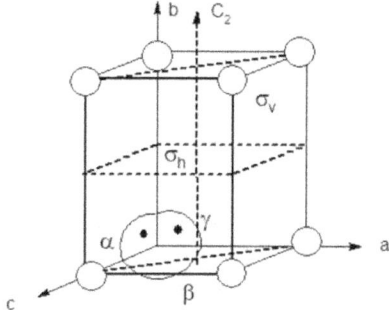

(the vertical mirror plane is equivalent to the C_2 rotation, so it is not mentioned in the Hermann-Mauguin notation)

| 35 | Unit cells: |

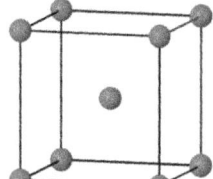

Lattice type: body centered I

Crystal system: cubic

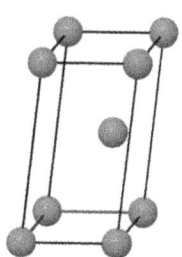

Lattice type: body centered I

Crystal system: orthorhombic